The Extraterrestrial Species Almanac

THE ULTIMATE GUIDE TO
Greys, Reptilians, Hybrids, and Nordics

CRAIG CAMPOBASSO

MUFON
Mutual UFO Network
— est. 1969 —

This edition first published in 2021 by MUFON, an imprint of
Red Wheel/Weiser, LLC
With offices at:
65 Parker Street, Suite 7
Newburyport, MA 01950
www.redwheelweiser.com

ISBN: 978-1-59003-304-3
Library of Congress Cataloging-in-Publication Data available upon request

Cover and text design by Kathryn Sky-Peck
Cover art © Kathryn Sky-Peck
Interior photos/images by Craig Campobasso; see also credits page 278
Typeset in Warnock Pro
Printed in the United States
IBI
10 9 8 7 6 5 4 3 2 1

Contents

For my biological father, Frederick Warner Vierow

My mother, Marie Donna King Campobasso, told me when I was twelve that my father, Louis Campobasso, adopted me when I was a year old, when they married. Mom then revealed to me what she knew about my biological father, Fred, whom she had lost track of. (It wasn't much.) Periodically throughout my life, I tried to find Fred with no success. In 2019, a genealogist found him for me and told me he had died in 2006. I found out where he was buried and retrieved a man's name from the mortuary who had called them when Fred passed. This man turned out to be Fred's half-brother, and he lived around the corner from me.

I met my newfound Uncle John and have begun a fantastic friendship with him and his family. I learned that my father was born in 1934 and was in the Air Force. Uncle John told me that Fred worked for Project Blue Book, the US Air Force study on unidentified flying objects that started in 1952 and investigated more than 11,000 sightings, until it was decommissioned in January of 1970. Its main study was to determine if UFOs were a threat to national security, and to scientifically analyze UFO-related data.

When Fred left the Air Force, he became a set builder at Warner Brothers and was credited on projects as the head of construction.

I don't think it's a coincidence that my father and I have so much in common. I am writing a book on extraterrestrials visiting Earth—who they are and what they want—and Fred worked for a government organization that studied who they are and what they want. Plus, I have worked in the entertainment industry casting film and television for the main part of my career; the bulk of Fred's career was in the film and television industry as well.

Like father, like son.

I believe there are no accidents in the universe. All is divinely planned.

• • •

Welcome to the multiverse. You are not alone.

Introduction

The idea that extraterrestrial life-forms exist fires up controversy. Everyone wants evidence, photographic and otherwise, to ground the claim in cold, hard fact. There's a saying: "For those who believe, no explanation is necessary; for those who do not, no explanation will suffice." Whether you are a believer or not, this book will thrust you into the unknown universe of benevolent and malevolent aliens. This almanac is a preparation guide, a bible of cosmic life, for the day when disclosure arrives, when global governments and religious factions tell the truth about the existence of extraterrestrials, and admit that ETs have been with mankind throughout history.

Assembled in this book are eighty-two extraterrestrial races that have interacted with all walks of life on our planet. You'll discover obscure, age-old, and new cases of personal contact, some ongoing and others random chance encounters. These firsthand accounts give us a cosmic window into where the extraterrestrials came from, what they look like, their belief systems, cosmic agendas, technologies, consciousness abilities, and dimensional capacity.

I have compiled a wealth of my favorite contactee cases, eyewitness testimony about beings from Alpha Centauri, Aquila, Cygnus Alpha, Itipura, Rigel, Sagitta, Taurus, Zeta Reticuli, and more, including a parallel Earth three hundred years in the future. This book presents a thumbnail review of these races and others, plus further reading suggestions for those interested in diving deeper on a particular species.

Perhaps the strongest evidence I offer here for the existence of extraterrestrials, besides accounts from researchers, experts, government officials, innocent bystanders, and ET enthusiasts, is a photograph I took with my Minolta camera while investigating an abduction case in the Carolinas. It's of an interdimensional being holding a light source and wearing eye goggles, perhaps used for interdimensional vision, like how we use night vision binoculars. I've also included photos taken by contactee Maurizio Cavallo—images of a male and female of the Clarion race who come from the Aquila Constellation. Also, to satiate your visual curiosity, portraits of the various cosmic species have been painstakingly brought to life.

While it seems that most of the universe is humanoid (and benevolent), there are also intelligent, malevolent life-forms as well, including reptiles, Greys, dinosaurian beings, insectoids, and even some humans. DNA is more precious than any alloy across the cosmos, and genetic engineering and manipulation are standard practice. Once a race becomes schooled in genetics, they might create stock for a variety of reasons. For example, our Earth has been speculated to be an experimental laboratory for extraterrestrials. Are these genetic masters uniting our DNA with extraterrestrial DNA, giving us a genetic upgrade? Are new lines of human evolution being created so that we may return to the cosmos as fully conscious beings, ready to participate in the next realities? Will we then become the tech gods of hybridization with younger races across the cosmos? The answers to these questions, and many more, are explored throughout the book.

Finally, appendixes on the Galacterian Alignment and ship technology, cosmic law, and otherworldly terminology help explain what lies beyond the cosmic curtain. Here you will discover the highly developed spiritual technologies incorporated into extraterrestrial spacecraft, as well as the cosmic laws and dimensions that regulate the universe. Although these futuristic concepts might challenge your belief system, one thing is for sure—your insatiable curiosity will stir, begging for more answers, especially to the greatest question of all: What is the meaning of life?

Once our world rejoins universal society, our insight into how these races think will be a great advantage to us. We'll know which races to trust, which to beware of, and how to proceed once cosmic integration has begun out in the open.

Craig Campobasso
Los Angeles, California

Part One

HUMAN

EXTRATERRESTRIAL

RACES

Saint Germaine, Ascended Master

Agarthans

Universal Origin: Although Agarthans are citizens of Earth, they originally came from the stars when the ancient civilizations of Lemuria (250,000 years ago) and Atlantis (128,000 years ago) were established.

The capital of Agartha is Shambhala, the greater, located beneath the Himalayas.

Posid is the Atlantian capital located beneath the Mato Grasso plateau in Brazil.

Telos is the Lemurian capital located beneath Mount Shasta in Northern California. When Lemurians first arrived on Earth they settled across America and named their settlement Telos. There are entrance portals around the town and mountain of Shasta. There are also visible entry points at the North and South Pole; these openings have a circumference of about 1,400 miles. No one may enter Agartha unless they are an appointed initiate.

Agartha is about three-quarters land and one-quarter water. There are oceans, lakes, rivers, and ponds. Fresh water springs provide drinking water. Agarthans employ advanced hydroponics for growing food. In the center of Agartha, the plant and animal life is much larger than what is found on Earth's surface. The light source is an inner nonblinding sun that is synchronized to the outer sun. When night comes, cloud cover travels over the rays of light and retracts upon the break of dawn. When the cloud phenomenon occurs, the sun takes on a faint red-golden glow. The skies are shades of purple to violet, but the real beauty comes from the hues that paint the clouds like a watercolor fresco.

In Telos, the central light (sun) is derived from a special crystal set high in the cave. When night arrives, filters dim the light, and when dawn arrives, the filters slowly peel away to reveal the progressing day.

Agarthan homes vary in design. In Telos, residences are made of smooth stone and are lit by light crystals. Other cities are Grecian style, with or without pillars, or adobe-like, and others are made of crystal. Temples are Grecian appearing with pillars and adorned with gemstones to enhance the structures' vibration, thereby elevating the consciousness of all who enter.

Advanced machines bore into earth to create transport and/or walking tunnels, the aftereffects ensuring strong walls coated with a flexible bonding material to maintain stability. When walking through darkened areas, such as tunnels, one's body heat activates the walls to glow with light.

Agartha is the cradle of our current civilization.

Physical Characteristics: Agarthans are human and mirror surface dwellers, except for their height. The average Agarthan is six to eight feet tall, while some can reach twenty feet.

Lemurians' skin tones are fair, while Atlantians are dark brown to fair-skinned with dark and reddish hair.

Belief System: Cosmic Law of One. We are all one and we serve One.

In Telos, Adama, an Ascended Melchizedek Master is the Lemurian High Priest along with the High Priestess Terra Ra. All Agarthans are trained in the Ancient Melchizedek and Angelic Corp Orders. Although a large number of Agarthans are ascended masters, others labor toward ascended status. The interior of Venus is the hub of the Melchizedek Order in our solar system.

Agarthans lives are long and their young usually do not start their formal education until the age of twenty to twenty-two but may start earlier depending on individual choice.

As adults they may choose up to five live-in relationships (bond marriages) in different households, their time divided equally. All forms of sexuality are embraced. Learning how to love unconditionally in personal and professional relationships is paramount. With up to five mates, they are able to seek their true inner feelings, express their hearts, and explore each relationship. All are in accordance with these arrangements, only wanting their mate to experience and express love at its full potential. Over the contemplation of time, ultimately each person will choose one person and enter into a sacred marriage. Children are not conceived in bond marriages, only in sacred marriages.

Cosmic Agenda: Agarthans participate in the Star Seed Alignment Program, raising consciousness on the surface. When Earth makes the shift from duality to cosmic consciousness, the civilizations above and below will become One Unified Earth. At that time Earth will change locations in the universe, having reached a higher dimensional coding, and enter into a sector closer to the central sun. Earth's current locale is sector nine, but will reach either sector two or three, depending on the evolution of our global consciousness. At that time, Agarthans will assist, along with other star nations, Earth's reintegration into universal society, as our planetary consciousness will then be vibrating at a much higher rate.

Agarthans are members of the Galacterian Alignment of Space Peoples and Planets, a universal alliance that works for the betterment of all universal kind.

Technology: Agarthans have Galacterian motherships, starships, thought-ships, fleet ships, and scout ships. All craft and super-computer systems are based on organic consciousness technology. Lemurian and Atlantian discus-shaped ships are called the "silver fleets." Their main base is north of Salt Lake City, Utah, at an undisclosed location.

Sleek bullet trains run on an antigravity force field and travel at unbelievably high speed. For instance, it would only take two hours to ride from Telos to Shambhala. During the smooth ride there is no sensation of passing through time and space.

The Galacterian Alignment's home base in our solar system is the interior of Saturn.

Consciousness Abilities: Agarthans are fully conscious, sovereign, and telepathic beings. They speak many star nation dialects, including the language of light. They also speak the sacred solar language— Solara Manu. Some languages, such as Sanskrit, Egyptian, Hebrew, and Ancient Celtic have inherited a few words from this language.

Dimensional Capacity: Multi-dimensional beings.

Saint Germaine and the I AM Teachings

In 1930, while hiking Mount Shasta in Northern California, Guy W. Ballard met Ascended Master Saint Germaine. It was through Saint Germaine's assistance that Ballard created a series of books on the Mighty I AM Presence (the God in you), a way to make contact with the Forces of Light.

Altarians

Universal Origin: Altarians are from the Aquila constellation. Inhabitable worlds in their immediate vicinity are home to other humanoids, reptilians, and Greys. Altarians are great cohabiters with other races.

Physical Characteristics: Altarians are indistinguishable from Earth humans, though they are taller: Men are six to nine feet. Women are five foot nine to seven feet. Altarians are a beautiful race; men have toned, muscular bodies, and women are voluptuous. Although most are light-skinned, some Altarians skin tones range from blue, green, tan, and brown, to a variety of other hues, from breeding with other human races. The skin tone spectrum is a natural occurrence within star nations.

Altarian's also study the colors in suns and stars in planetary systems to understand how the frequency of light denotes our different skin pigmentations.

Belief System: Altarians are a very intelligent, wise, and learned race, and they follow the Cosmic Law of One. They study science and natural philosophy. It is in their nature to analyze fundamental questions about creation, to study other human societies from across the universe, and note the similarities and differences due to culture and environment.

Altarians believe that the universe is a symphony and we are its instruments. As we learn from planet to planet, raising our octave, our frequency, our soul's song, that in divine time, we will ultimately master the whole composition.

Cosmic Agenda: Altarians are members of the Galacterian Alignment and participate in the Melchizedek Star Seed Program. The

Altarian Female

Altarians will become our cosmic teachers once we are a fully conscious race.

In the fifties, Altarians reached out to US government officials and asked them to disclose their existence and of life in the universe. The government declined.

Technology: Altarians have Galacterian motherships, starships, thought-ships, fleet ships, and scout ships. All their craft and command centers use biological supercomputers. All craft are able to travel through time-space funnels and the innerspace continuum.

Their home world's modes of transportation are teleportation portals designed for one or groups of any size. They also have sky buses that run on magnetic currents.

Consciousness Abilities: Altarians are fully conscious and sovereign beings. They are sympathetic telepaths; they feel a profound depth of others' soul history and recognize the beauty of it.

Altarians explore the cosmos and dimensional layers by astral traveling in groups. It leads them to territories in their own dimension they wish to study and explore.

Dimensional Capacity: Altarians have multirealm commandership. They are able to reassemble their physical bodies or any matter from one location to another through a molecular stimulation of light codes. They can sweep an entire civilization from the ground up into a mothership in less than a second.

Some Altarians come to our space-time from the future.

Andromedan Being

Andromedans

Universal Origin: Andromedans are so named because they come from the Andromeda star system.

Physical Characteristics: Andromedans are blue hairless humans with three classifications: female, male, and androgynous. Females are five six to six feet; males are five nine to six six; androgynous beings are in the height range of the males and females. Andromedans are a beautiful classification of beings; their hearts actually feel like heavenly stars are alive within every chamber, magnetizing all they come in contact with, to experience true nirvana.

Belief System: Andromedans believe in the Cosmic Law of One. As a member of the Galacterian Alignment, Andromedans are at the pinnacle of unconditional love and council the star nations. The twenty-four elders of the throne are located in the great city of Beta Andromedae in Andromeda. Master teachers and constellation fathers from every star nation are counseled by them to mend problems under their outer space jurisdiction.

Cosmic Agenda: Andromedans endeavor to embrace all life with grace, dignity, and love; to share cosmic knowledge and wisdom; and to assist souls ready to advance into the grandeur and tiers of the universal consciousness. Their culture embraces all classes of master teachers either attained through ascension or created status.

Andromedans participate in the Melchizedek Star Seed Program. When Earth makes the dimensional shift into full consciousness, they, as well as other Galacterians, will become our cosmic teachers, our supreme mentors of knowledge and wisdom. They are sought after because of their diplomacy with all worlds.

Technology: Andromedans have Galacterian motherships, starships, thought-ships, fleet ships, and scout ships. All their craft and command centers use biological supercomputers. All craft are able to travel through time-space funnels and the innerspace continuum. Andromedans have galactic command centers stationed above and below the Earth.

The ancient masters of their civilization travel in Merkabah vehicles. They create a transport around themselves to either travel short distances or enter star towers (space portals) to journey to other sectors of the universe. The latter takes place in their great city of Beta Andromadae.

Consciousness Abilities: Andromedans are fully conscious and sovereign beings, and are one of the most spiritually advanced races in our universe. They speak the light essence language. They are also capable of holographic telepathy: using the third eye as a projector to screen their personal story for others to watch in 5D technology, a depth perception so real, it has to be seen to be believed.

Dimensional Capacity: This race has multirealm commandership. They are able to reassemble their physical bodies or any matter from one location to another through a molecular stimulation of light codes. They can sweep an entire civilization from the ground up into a mothership in less than a second.

Some Andromedan races come to our space-time from the future.

Twenty-Four Elders
of Andromeda

Universal Origin: The elders of Andromeda reside on Callonus Planet in the great city of Beta Andromedae, also known as Merkabah Star Station 144. They are sovereign to twenty-four vast territories that monitor 144,000 territories, of which Earth and the Pleiades star clusters are a part. The Great Star Station boasts a far-futuristic garden of star towers, portals connecting to portals across the universe, a port of call for master teachers when visiting the elders.

Physical Characteristics: The elders of Andromeda, twelve women and twelve men, are human celestial gods and goddesses. They have an array of skin tones. Some have hair and others are hairless.

Belief System: The elders of Andromeda are pure divine consciousness. They govern Living Light programs and distribute them to the Angelic Corp to disperse throughout Creation and to younger sectors for new creations. They are the divine scribes of the Cosmic Law of One.

Cosmic Agenda: The elders of Andromeda are the pinnacle of governing bodies in the universe. They mentor ascended and created status master teachers in all areas of life that pertain to their constellation or planetary sphere. They are currently focused on Earth's consciousness-raising programs, and work closely with the Galacterian and Star Seed Alignments.

Technology: The elders of Andromeda travel in Merkabah vehicles. They create a transport around themselves to either travel short distances or enter star towers (space portals) to journey to other

Elders Elizolet and Arramann, the Key Speakers

sectors of the universe. The latter takes place in their great city of Beta Andromadae.

Consciousness Abilities: The elders of Andromeda have avatar and solar consciousness. They speak the light essence language. Due to their elevated spiritual stature, they use advanced telepathic channels known as *coagular wave bending* to send messages across the universe, or through time streams, unlike the regular limitations of telepathic distances.

Dimensional Capacity: The elders of Andromeda are multidimensional. They are able to wear a dimensional layer like a garment.

Archangel Michael

Angelic Corps

Universal Origin: The Angelic Corps is from the Isle of Paradise, located in the center of all creation.

Physical Characteristics: The Angelic Corps is a multiplicity of universally created beings whose physical appearances mirror cultures from across the universe.

Belief System: Angels of all classes work for the Source of All Existence, for the Invisible Architect of Creation, for God, no matter what your belief is. Every being was given a body to perform right action to expand their spark from the formless to form. Seraphim, for instance, guide man along the pathway of experiential life to increase soul growth, to ultimately transverse and ascend from learning planet to learning planet. Accepting angelic guidance is a brave feat for the student, as it means life will be without ease. But to choose to climb a steep mountain with the lesson at hand, and then reaching the mountain top, is gratifying for the soul. Confidence is built and spiritual rewards are plentiful. Mortals are also assigned guardian angels, who oversee the oversoul's journey and assist from the unseen realms. Ministering angels minister daily to the individual from behind the veil.

Cosmic Agenda: The Angelic Corps maintains the calibration and Living Light across the universe, and also assists its creatures by delivering curriculums of Living Light to worlds ready to rise up and out of duality.

The Angelic Corps oversees the Galacterian Alignment. They labor in unison with the University of Melchizedek in the Mira System, the realm where Living Light programs are dispersed.

Technology: The Angelic Corps are the architects of advanced spiritual technology and incorporate them into worlds ready for an upgrade.

Consciousness Abilities: The Angelic Corps speaks the light essence language. Angels (created beings) have full divine power and will. They can think anything into existence, teleport anywhere, heal with a thought, bring back the dead with a touch, and send a telepathic or holographic message across the superuniverses. Their divinity, their power, and their unconditional love are limitless.

Created beings *are not* born of man and woman but created from the godhead. They do not have belly buttons. Paradise Sons such as Archangel Michael and other archangels, angels, gods, and goddesses, all fall under this category. Some created beings are created as adults, while others are created as infants and assigned to dedicated, celestial parents to rear. It takes special parents to raise a created being. The children's advanced consciousness and power require proper development and nurturing.

Dimensional Capacity: The Angelic Corps, dependent on their classification, can enter multiple dimensions, universes, or superuniverses. Once they are assigned to a specific universe, they will stay until their mission is complete, although they may travel through space and time to gather information relevant to their tasks.

Antarians

Universal Origin: Antarians hail from Antares, or Alpha Scorpii, the bright red star marking the scorpion's heart in the Scorpius constellation.

Physical Characteristics: Antarians are indistinguishable from Earth humans, though they are a bit taller. Women are six feet to seven five, and men are seven to nine feet. Their skin tone ranges from white to olive, brown, or a unique copper color. They are a fit race and quite attractive.

Belief System: Antarians follow the Cosmic Law of One. They believe we are mirrors of each other, and therefore they treat all they come in contact with in honor, respect, and love. Their ability to experience and peer into the window of every soul, to understand their soul history, is humbling to them. They believe the universe is our master teacher and we are its students; that our consciousness will ultimately become identical to the universe and merge with it. But while they/we elevate up through the dimensions, they/we become the master teachers to younger races. There is no threshold in learning; every being is constantly evolving.

Cosmic Agenda: Antarians are part of the Galacterian Alignment and participate in the Melchizedek Star Seed Program, which is currently focused on raising consciousness on Earth and preparing for a dimensional shift. Once Earth beings are fully conscious, they, as well as other Galacterians, will become our cosmic teachers.

The Antarians assist from a higher realm, the elevated dimensions—their home, stepping down their energy to interact with humans about life elsewhere. They assist star seeds in their spiritual development through dream states, downloads into consciousness, and astral visitation.

Antarian Male

Technology: Antarians have access to Galacterian motherships, starships, thought-ships, fleet ships, and scout ships. All their craft and command centers use biological supercomputers. All craft are able to travel through time-space funnels and the innerspace continuum.

Antarians, like other star nations, have access to the universes' creational codes. These principles are incorporated into general Galacterian Alignment technology. Star nations tweak the tech based on their collective preference. The Antarians, for instance, incorporate the healing properties in light, tuning its frequency, sound, and color to higher realms, and illuminating the silver metallic walls of their craft with it. The walls are interactive and connect to the ship's brain. All an Antarian has to do is connect their thoughts to it and create whatever their desire is at will. From food to clothes to travel within the ship and across the universal divides, Antarian space vessels create a transcendental experience.

Consciousness Abilities: Antarians are fully conscious, sovereign, and telepathic beings. Living in the higher realms, some Antarians are light beings or are a blend of spirit and physicality. They are able to seamlessly interact between densities.

Antarians assist star seeds with diseased areas in the body, mental overloads, and belief system collapses as spiritual elevation increases. They focus light through their third eye and can heal instantly through mental focusing.

Dimensional Capacity: Multirealm commandership. Antarians are able to reassemble their physical bodies or any matter from one location to another through a molecular stimulation of light codes. They can sweep an entire civilization from the ground up into a mothership in less than a second.

Some Antarians come to our space-time from the future; they astral travel to visit star seeds on Earth in the third dimension.

Anunnaki Male

Anunnaki

Universal Origin: The Anunnaki are an old generation of gods from planet Nibiru, which in the Sumerian dictionary translates to Jupiter. The name Anunnaki is taken from An, the Sumerian god of the sky.

Physical Characteristics: The Anunnaki are giants compared to man today, with olive complexions and dark eyes. The men wear their hair long in soft waves, with long beards curled in rows, coiffed to perfection. Women have long, wavy hair and dress similar to Grecian goddesses. They possess superhuman strength and have powerful presences.

Belief System: The Anunnaki believe in service to self. They have been convicted of breaking many cosmic laws, including creating a slave race on Earth through genetic manipulation.

Cosmic Agenda: Four hundred and fifty thousand years ago, the Anunnaki came to Earth in search of gold, an element needed to mend their fragile atmosphere. Nibiru, their world, was endangered from radiation from their sun. They settled in Sumeria. Through genetic manipulation, they created a slave race to mine gold. Back home they transformed the gold into a liquid light modification and blanketed it around their atmosphere to shield the harmful sunlight. Eventually, they left Earth with their lower vibrational gene pool continuing to evolve.

It is said the Anunnaki will return to Earth; for the cosmic law of cause and effect is growth over time, and they are responsible for their human creations. Others say the Anunnaki are already here and maintain control over human evolution through covert elite groups.

Some say the Anunnaki are human reptilians or reptile-influenced. The Draconians and Anunnaki are said to vie for the dominion of Earth. Or are the Draconians and Anunnaki one and the same?

Today, extraterrestrial enthusiasts say the Anunnaki have finally come to terms with the errors of their ways and are making every effort to correct their karma.

Technology: The Anunnaki have great spacers—motherships designed with multiple cities, replicas of their past civilizations. Their fleet craft are two half-domes that form an oval shape. When in transport mode, tiny organic feelers extract from the center of the ship to sense the universal vibrations. Once the ship has found its resonance octave, it propels itself into the antimatter-verse, into old secret hyperspace corridors created by their ancestors long ago, for instant teleportation across the universal divides. They prefer to travel incognito to avoid detection from hyperalert Galacterian Alignment eyes.

Consciousness Abilities: The Anunnaki enjoy created being status. Their divine power has extensively diminished over time due to their egos. But what they used to be able to do with the power of their minds, they are slowly replacing with technology.

Dimensional Capacity: They're capable of living and traveling up to the sixth realm.

Apunians

Universal Origin: Apunians hail from Alpha Centauri, one of the brightest stars in the southern sky, which is in the Centaurus constellation. The first reports of these human extraterrestrials were in the Peruvian Andes in the early fifties. The inhabitants of the Andes, the Quechua people, came in contact with the Alpha Centaurians and named them Apunians in honor of the mountains. *Apu* in Quechua means "Lord, protector, and spirit of the mountains."

Physical Characteristics: Apunians are an attractive, Nordic-looking race and are basically indistinguishable from Earth humans except they are taller. Females are six foot five to nine feet. Males are six foot nine to twelve feet.

Belief System: Apunians follow the Cosmic Law of One. They are aware of the higher self/oversoul, and of the transcendental reality that unifies the whole of Creation. Apunians say that there are 1,019,000 civilizations in the universe.

Cosmic Agenda: They are masters at transforming negative energy. They embrace complete service to the citizens of all worlds, and consider themselves brothers and sisters to all. They are protectors of younger races and respectful of all life, and care for all sentient life equally. These words are not a part of their vocabulary: *advantage, favoritism, preferences, privilege*. The Apunians feel a deep connection to assist us in raising our vibration into a higher consciousness, as they have merged their own consciousness with their higher self, to ultimately gain full access into universally awake and aware consciousness. They have been visiting our world for a very long time, choosing heart-based contactees to spread the word of their existence and message of brotherly love.

Top: Apunian Male; Bottom: Apunian Female

Apunians assist the poor regions in Peru by healing their sick and making it rain for their crops, among other humanitarian deeds. Apunian contactee and author Ricardo Gonzalez clarifies that they don't overstep, though, saying, "The extraterrestrials may support us, but they will never solve our problems. The real contact is with ourselves."

Technology: Apunians have an impressive assortment of Galacterian motherships, starships, thought-ships, fleet ships, and scout ships. All their craft and command centers use biological supercomputers. Other ships are shaped like winged birds, butterflies, cylinders, pears, and clover leafs. Their display screens can go backward and forward in time, projecting the historical past of any planet or person who stands before the screen, including images of a person's thoughts.

Apunians can contact other worlds through an artificially created space-time bubble they call a Xendra. The Xendra is where their physical reality and our physical reality converge to meet and converse. Think of it as a half-orbed light portal above ground between two worlds.

Apunians move around on their planet through a levitation apparatus connected to their belt, ankles, and wrists. The device is teeming with positive ions, and when in use their weight condenses, allowing them to move at the required speed. They can zip across the air, stop on a dime, move horizontally or vertically, elevate or descend.

Consciousness Abilities: Apunians are fully conscious, sovereign, and telepathic beings.

They can heal the sick by condensing the smallest particle between the void and the material, and then regenerate the cells to perfect harmony and health. They have attained near-perfect immortality by possessing the knowledge of deconstructing the atom to its smallest parts, and having command of positive ions. They protect and enhance all cellular life whenever permissible and possible.

Apunians have the innate ability to control nature. They are also telekinetic, able to move objects with the power of their mind. They are levitation experts when building their cities.

One of their greatest traits that they hold dear is that they transfer love to one another daily, reinforcing the human bond, ensuring happiness for the entire population.

Dimensional Capacity: Apunians are multidimensional beings. They are able to reassemble their physical bodies or any matter from one location to another through a molecular stimulation of light codes.

Some Alpha Centauri races, like the Apunians, come to our space-time from the future.

Arcturians

Universal Origin: Arcturians come from Arcturus, a red giant star in the northern hemisphere of Earth's sky and the brightest star in the constellation Boötes (the herdsman).

Physical Characteristics: Male and female Arcturians appear somewhat uniform, with slight variations in features, and heights from ten to twelve feet. Their bodies are thin, with lanky arms and legs. When they walk, they glide with elegance. Their skin is either blue or milk white. They have large craniums with small noses and mouths. Arcturians in higher density forms are that of light and consciousness. Depending on what density they are from, their appearance alters. The higher the density, the more they become light.

Their large, almond-shaped eyes are human in appearance and run the color spectrum. They have a pair of nictitating blue lenses that can cover the whole eye, which act as a light filter and protective barrier, like sunglasses.

Belief System: Arcturians believe in the Cosmic Law of One. They serve all universal-kind as stewards of love and light, of right use of will, of merging dualistic natures to inner peace. They share with willing civilizations how to integrate their forward-thinking, proven belief systems and emotional stability to resolve conflicts. Edgar Cayce said that Arcturus is one of the most advanced civilizations in the galaxy. Arcturians serve the Supreme Creative Force and protect the sacredness of life throughout creation. They are governed by a council of elders.

Cosmic Agenda: Arcturians are part of the Galacterian Alignment and participate in the Melchizedek Star Seed Program. Arcturian star seeds on Earth use holographic healing to assist in creating templates for fifth dimensional and higher dimensional experience

Arcturian Female

and ascension. They assist in attuning the heart and mind back to a natural resonance frequency of light, which also helps ease ascension symptoms.

Arcturians have galactic command centers stationed above and below the Earth. They work closely with all brotherhoods and sisterhoods of light, and the galactic commands within the Galacterian Alignment. They are guardians of advanced consciousness in this universal theater and grant all creatures access to their knowledge and wisdom.

They are successful diplomats between star nations, and have a serene, calming effect when speaking and/or visualizing imagery to reveal how both points of view will be able to cohabitate peacefully. After suggestions are made to mend wounds, the Arcturians bring to light future visions of a harmonious outcome, which is seen and felt deeply by both parties, and is the catalyst for them to make amends. Arcturian supreme healers are known as the heart of the cosmos for their philanthropy.

JJ Hurtak, in *The Book of Knowledge: The Keys of Enoch*, describes Arcturus as a midway station (a programming center) used by the physical brotherhoods of light to oversee the many sets of experiments with "physical beings." They are the designers of some of the crop circles on Earth, encoding new frequencies into the planet and in her creatures.

Technology: Arcturians use Galacterian motherships, starships, thought-ships, fleet ships, and scout ships. All craft are able to travel through time-space funnels and the innerspace continuum.

Their technological advancements are due to advanced frequencies derived from supreme spiritual elevation. Although they use Galacterian craft, all star nations have variations of technology. Arcturian starships and saucer craft are unique, as they are fueled by light, sacred geometry, and crystalline energy powered by the central sun. Their geometrical navigational calculations are what move them through space and time and other realties. They are master sound and celestial healers and sacred geometry experts, and are versed in color therapy as well.

Consciousness Abilities: Arcturians are fully conscious and sovereign beings. They use the light essence language. They are holographic and sympathetic telepaths, and are able to access coagular wave bending to send messages across the universe or to another, or through time streams, unlike the regular limitations of telepathic distances. They possess the power of telekinesis. Some come to our space-time from the future.

Dimensional Capacity: Arcturian fifth- and sixth-dimensional beings primarily work with Earth's consciousness-raising programs. They interact with light beings of their own race in the elevated dimensions and other star nations' collective consciousnesses from the ninth and tenth dimensions.

Arcturians have multirealm commandership. They are able to reassemble their physical bodies or any matter from one location to another through a molecular stimulation of light codes. They can sweep an entire civilization from the ground up into a mothership in less than a second.

Arians

Universal Origin: Arians come from the planet Arian, located in the Aldebaran system in the Taurus constellation.

Physical Characteristics: Arians are indistinguishable from Asian races on Earth. Women and men are slender and stand from five foot nine to six foot six. Their faces are oval-shaped with attractive features, and framed by lustrous, black hair. Their eyes are usually brown or black, and skin tones range from golden to yellow to tan. Their life spans are over four thousand years.

There is another Asian race, called the Gne, from planet Pall, in a neighboring system that is a part of their crew. The only difference in appearance is they have small, bud-shaped noses.

Belief System: Arians believe in the Cosmic Law of One. They believe all souls are comprised of the primordial cell, and because of this all are one and the same. Their advanced consciousness allows them awareness of the universe's invisible and visible realms. They thrive on working with other races collectively, enjoying the bonds of universal brotherhood.

Cosmic Agenda: Arians are explorers of the galaxies, and their members include anthropologists, psychologists, biologists, geneticists, and gene specialists. They are experts of intergalactic languages—spoken word or telepathic. They study men and women from various civilizations across the universe, and collect genetic materials. When they come in contact with younger civilizations, such as Earth, they share their civilization's history when they were at war with other worlds, and how they overcame their challenges. They warred for so long, and there was such a loss of life, that after countless years both sides came to realize how useless war was. They stopped manufacturing weapons

Arian Female

and came together. They evolved, explored space, met their planetary neighbors, and built a cosmic community.

On worlds in conflict an emergency Melchizedek son is dispersed to be an intermediary between both sides. These sons function on many diverse levels of the universe, in this and other dimensional layers. They are able to do the impossible. One might think they are the very breath of God when in their presence.

Technology: Arian fleet craft use two fusion engines that operate in a vacuum, allowing them to travel through the antimatter universe. It takes two weeks to warm up their far-futuristic engines when they have been sitting for a time. Their elongated motherships are based on Galacterian technology and have extending appendages in the front, middle, and back. From tip to stern, the craft spans about 2,691 feet. They have eleven other races onboard their mothership from Aldebaran or neighboring systems.

Being language specialists, they have a translator box for every galactic language they have come across.

Arian healing machines are able to separate the physical and spiritual bodies, and then slice the pertinent elements for viewing. Once the source of a disease is isolated, colors and sounds are then employed for instant healing.

They are able to grow replacement limbs that function like natural limbs by using the patient's own DNA.

They're also genetically engineering new animal species that have a symbiotic relationship to the environment they are placed in.

Consciousness Abilities: Arians are fully conscious and sovereign beings. Their telepathic headgear amplifies brain frequencies 100,000 times to open channels of communication to the biological supercomputer onboard their craft.

Dimensional Capacity: Arians are multidimensional. They use the antimatter universe for travel, cutting travel time to a minimum.

Cassiopeian Male

Cassiopeians

Universal Origin: Cassiopeians originated in the Cassiopeia constellation.

Physical Characteristics: Cassiopeians are indistinguishable from Earth humans except for clear webbing between their fingers that's unnoticeable unless their fingers are spread apart. This feature developed over time, as they are avid swimmers and lovers of ocean life. Their slim bodies skim through water the way dolphins glide across the surf.

Belief System: Cassiopeians believe in the Cosmic Law of One. Water is sacred to them and is purified by the created beings on their world and stored in clear quartz crystal vats. Goddesses meditate over the vats to infuse the liquid with love. The Cassiopeians' devout practice in water purification has greatly benefited them internally and externally. Internally, the purified water has activated a crystalline matrix within the blood cells that gives them endurance unlike any other species. Its influence on the brain's cellular structure has expanded the capacity of storage beyond 100 percent and has categorized them as superhumans.

Cosmic Agenda: Cassiopeians are part of the Galacterian Alignment and participate in the Melchizedek Star Seed Program.

They study oceans on their world and on other worlds. Communication with sea creatures is either through telepathic static or moving pictures, or by telepathically speaking their language through a mind-extension apparatus.

The Cassiopeians are looking forward to meeting their extended family on Earth. Many from their system are stationed in galactic command centers above or below our planet. They are excited when measuring the planetary consciousness as a whole and bear witness to our spiritual elevation. Star seeds from Cassiopeia are usually focused on environmental issues, especially ridding the oceans of pollution.

Technology: Cassiopeians have Galacterian motherships, starships, thought-ships, fleet ships, and scout ships. They also have specialized water ships similar to the spiritual engineering of a Merkabah ship, based on liquid light principles. Riding in one of these craft makes one feel light as a feather, and mind clarity increases one thousand-fold. It's likened to nirvana.

Their thought-ships, fleet ships, and scout ships are just as agile in the water as in the air, in complete control of the elements. The Cassiopeians helped create underwater technology that is used throughout the Galactic Kingdom. Upon entry into any body of water, a slippery cocoon of light surrounds the craft, the protective barrier that allows the ship to slide through liquid with ease. It is also designed to not disturb ocean life.

All their craft and command centers use water-based biological supercomputers, as if they were living creatures from the sea. All craft are able to travel through time-space funnels and the innerspace continuum. They are developing a wormhole transport made from liquid light, a fluid infused with divine consciousness, capable of bending space to other universes.

Their healing tools are water- and crystalline-based. With their heightened mental clarity, they astral travel to higher realms and bring back sacred light, brilliant opalescent colors, and infuse their healing wands, beds, and chambers with this light in weightless liquid form.

Consciousness Abilities: Cassiopeians are fully conscious and sovereign beings. Besides being astute telepaths, they have an extra ability unique to their race: When speaking mind to mind, there is the usual attachment to feelings of the experience—but another dimension is added, a conscious union to the Living Library of Akasha that enhances the story with the amount of spiritual elevation gleaned from the sender's lesson at hand. The receiver also elevates, almost half of the sender, just by being brought into the experience on an unfathomable depth of soul connectivity.

Dimensional Capacity: Cassiopeians have multirealm commandership. The Cassiopeians are experts at quantum physics. They understand that like the rings inside a tree trunk that reveal its age, the universal dimensional folds go back to the beginning of time. Dimensions in this universal theater are endless, just like the multiverse creation.

Ceitan Male

Ceitans

Universal Origin: Ceitans are from the Cetus constellation.

Physical Characteristics: Ceitans are indistinguishable from Earth humans. Women range from five five to seven feet, and men are from six to eight feet. Their race resembles Spanish, Italian, and American Indian heritages on Earth. Skin colors range from olive to brown, tan, and reddish-brown. Most of the Ceitan population are dark- or brown-haired. Although they are a fit race, some bodies are stockier than others. They have the same range of eye colors on Earth, but most common on their world is royal blue.

Belief System: Ceitans believe in the Cosmic Law of One.

Part of the star nation code of ethics is to live by natural law. When one achieves *direct knowing* that they are a cocreator with the Creation, their sovereignty in the cosmos begins in a spark of light. Ceitans vividly recall when their world was activated to full consciousness, and the sublime feeling that came rushing into their hearts from the universal masses that assisted them. They wholeheartedly stand by us as we transition to where they are consciously so we will come to know that feeling as well. The Ceitans are very heart-oriented, choosing love to reveal wisdom, allowing the inner spirit to guide them through life.

Cosmic Agenda: The Ceitans believe, as all star nations do, that spiritual connectedness to the universe is the foundation of brotherly love, the glue that binds every star nation. As a galactic community, and as part of the Galacterian Alignment, they wish to share all they have learned with younger worlds. Some of these teachings are released to Ceitan star seeds on Earth so they will make public who the Ceitans are: cosmic guidance counselors, our cosmic neighbors.

The Ceitans also release higher realm information into the global field of truth and trust. Think of this field as clear matter adorned with stars, and when it comes down over the physical body, the energetic circuitry, the thought forms, interact with individual consciousnesses to bear fruitful ideas through intuition.

Technology: Ceitans use Galacterian motherships, starships, thoughtships, fleet ships, and scout ships. All their craft and command centers use biological supercomputers. They have learned how to harness cosmology—the study of quasars and their pulses, which spans along with the rotation of the neutron star—and incorporate it into their propulsion drives, allowing them to explore the interior of black holes.

Consciousness Abilities: Ceitans are fully conscious and sovereign beings. Their telepathic abilities are connected to their heart center, and heightened feelings are transferred with conversation through sensory levels of advanced frequency consciousness. It is a cellular telepathic connection that allows a window into the manifold body overlays of the conversationalists. Look at it as recognizing your true self in all its glory in another self and vice versa.

Dimensional Capacity: Ceitans have multirealm commandership. They are able to reassemble their physical bodies or any matter from one location to another through a molecular stimulation of light codes. They can sweep an entire civilization from one location to another.

Ceitans work from the sixth dimension on down into third densities. They align themselves to the spiritual intelligence of each dimension. They are assisting Earth's subtle reverberation into the next density.

Celestials

Universal Origin: Celestials ascended from the astral realm. They are now joined within the universal firmament; their heart and mind consciously a part of everything in the seen and unseen worlds. They are celestial realm creators.

Physical Characteristics: Celestials are able to choose any bodily form from their past incarnations, or a combination thereof, from the physical or astral realms. They are inwardly beautiful and exceptionally good-looking in this rarefied state. Celestial garments are made from ultradimensional fabrics depicting scenes from the cosmos: stars, suns, moons, galaxies, nebulas, planets. A celestial fragrance emanates from their presence.

Belief System: Celestials are devotional beings to the Creation and to its creatures. Celestials are the overseers of the Cosmic Law of One. They work from behind the veil of the celestial curtain.

Cosmic Agenda: Celestials prepare written code comprised within Living Light and disperse these learning programs across the heavens. These blankets of awareness, ascension grids, guide a planetary consciousness to the next level. They work in union with the Angelic Corps and the University of Melchizedek. Some Celestials are in a class all their own.

Technology: When traveling the multiverse, Celestials create an express space vehicle from gammamyne, a buoyant milky-white matter infused with star energy, sunlight, or moonlight and high vibratory crystals mined from across their lands. They have full cognition of every living cell of consciousness in the firmament of their realm, and with a mere thought, the craft molds around them instantly. They

Celestial Male

can modify the craft while in flight, change its form—anything the mind can think of, they can achieve.

All of their light technology is derived from gammamyne and can be modified depending on the mission, with the infusion of other properties in the celestial heavens such as stardust, sunlight, moonlight, etc.

Consciousness Abilities: Celestials have avatar and solar consciousness. They speak the light essence language. There is consciousness in all light, and the Celestials program the heavens through light.

Dimensional Capacity: Being creators in the multidimensional realms, Celestials continue their evolutionary journey, ultimately ascending into the Paradise Spheres, worlds that are beyond sublime, beyond human description. Looking back on their lineage, for instance, if it was on Earth, an ascended Celestial might oversee their bloodline and share enlightenment though the dream state, a visitation, or from behind the veil. Some Celestials are created beings. They work in unison with ascended Celestials.

Clarion Photos © Maurizio Cavallo Jhlos

Clarion Male and Female

Clarions

Universal Origin: Clarions are from the planet Clarion, which revolves around two suns in a binary system in the constellation Aquila.

Physical Characteristics: Clarions are indistinguishable from Earth humans. Females on average are five five and males are five nine. They have petite, round facial features, various colors of hair, and angelic eyes (the same display of colors on Earth) that are slanted and almond-shaped.

Belief System: Clarions believe in the Cosmic Law of One. They believe in creating Light, in honor of the One with no name, the invisible architect of all Creation. They believe time and space are their temples, planets and stars their cathedrals, and the universe(s) their sanctuary.

They are aware that they (and all) sentient beings own the consciousness of the primordial atom and of eternity, and know that death is only transformational. The soul is eternal. They also believe the Invisible Architect of Creation fulfills destiny, and the celestial creators weave fate.

They believe souls evolve from planetary sphere to planetary sphere (universal schools) until they reach the consciousness of the Primordial One, of the Architect with no name. Until they reach this state of consciousness, they will be born into this universal theater over and over again.

Cosmic Agenda: Clarions are here assisting Earth in rising up and out of duality and into universal ascension; vibrating into the next dimensions. Some have been living on Earth for more than eighty years. One of their bases is shared with other intergalactic beings located below the center of the Amazon jungle.

Technology: Clarions have common discus craft that are about eighty feet in diameter. The metal outer part of the ship can become transparent, revealing the inner parts, or the outside environment. The interior of the craft has the appearance of infinity. They are able to contract or expand the space—the laws of physics are obliterated. The craft is a living metallic creature.

Via thought projection, standing in the center of the ship, a black cube rises from the ground to five feet; its upper surface is dark blue and its side has an oval cavity in its base. When the cube is touched, a large golden edge ejects from it, the key to the ship's guidance system. By touching the key, the route officer, or "onhotima" in Clarion, and/or the pilot is able to communicate with the organic brain of the craft. The cube then becomes transparent and showcases images. These images are transferred to the monitors and display graphs and star maps. Through the route officer's mind projections, the images become fluid and they then chart a course to their destination.

The Clarions have a frequency alternator that projects into space-time, overthrowing all laws of physics. The small round handheld sphere alters the atomic frequency of matter (living organisms) and purifies the cells. In other words, the sphere can relocate one to other dimensions, made possible by a controlled acceleration of a person's molecules.

Human DNA has a programmed syntax that can be modified as it reacts to modulated impulses. They can introduce colors, words, and music, and change the key in your genetic code, reprogramming them and causing them to react to the new dynamic sequences. Through sound and color dynamics they can reprogram cells. Information contained in DNA can be transmitted through words because languages possess the codified memories of the entire universe.

Clarions are masters of biogenesis and astral biology.

Finally, they have a touch-contact metallic surface on a piano-like instrument that when played with the fingers creates a pure energy cosmic composition.

Consciousness Abilities: Clarions are fully conscious, sovereign, and telepathic (mental synchronicity) beings. They are masters of physics, metaphysics, and ultraphysics—all facets pertaining to the visible and invisible equation.

Dimensional Capacity: Clarions believe there is no difference between physical reality and immaterial dimensions. They are able to traverse interdimensionally.

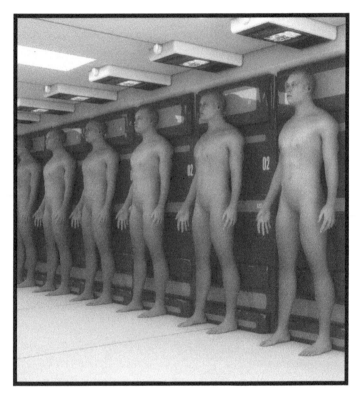

Clone Creatures

Clones

Universal Origin: Clones are biologically created.

Physical Characteristics: Clones could be human or alien; they're grown in artificial embryo sacs and void of a soul. (Souls are born of the Creation with divine love and intent.)

Belief System: Clones have no belief system; they are constituted consciousness units.

Cosmic Agenda: Clones are an extension of their programmers and can be used for good or evil purposes.

If clones are heavily implanted and joined with a group mind, either artificial or advanced alien, they become robotoids. The dark Orion (Draconian) Empire creates, uses, and sells robotoids. To protect themselves, Orions (Draconians) use robotoids to do their dirty work.

One type of clone exists to throw star seeds off their missions—interfering with alignment and ascension messages, preventing them from achieving spiritual elevation, and distorting the true benevolent cosmic reality. These clones use artificial intelligence consciousness to impersonate a voice of truth, but their messages are designed to create fear through false love. But as the star seed increases their spiritual sustenance, they are able to recognize the false voice and uproot it from their consciousness and detour it from future transmissions.

The first phase is recognition.

The second phase is mentally removing it from the mind through meditation by consciously creating a metallic sphere, a powerful magnet, and then turning on the magnet above the head, allowing it to pull all those negative and untrustworthy thought forms up into the sphere. Once the mind is cleansed, add pink (for love), purple (for spirituality), gold and silver (for Mother Father God) around the sphere. Let the sphere rise high into the sky and dissipate, allowing the colors and their meanings to transform the sphere and its contents.

The third phase is to create protection around the energy field by mentally focusing the same colors around yourself, and saying, "All that is not of enlightened universal consciousness (or God/Christ/Buddha/etc.) cannot penetrate this shield."

Soul replacement cloning is the dark art of partially stripping energies from a person's aura and replacing it with black energy to influence the person in many negative ways.

Technology: Some clones have animal souls, a spirit-capture technology taught to Earth scientists by aliens. Clones can also be controlled via implants in the physical brain or via holographic consciousness inserts.

Some clones are created through cattle DNA mixed with human DNA to create a near-human being. Consciousness is then withdrawn from a human and transferred to the clone through a computer to stimulate memories. The near humans are then bred with real humans to create Type 2 almost humans.

Consciousness Abilities: A clone's consciousness ability is limited; a soul matrix is only privy to those born of man and woman or to created beings. Clones have to be taught how to be human and interact with society. They are awkward and clumsy until they've reached maturity. They cannot attain a high IQ, and are easily manipulated by their creators.

Dimensional Capacity: Clones can only exist in the third dimension.

Created Beings

Universal Origin: Created beings are not born of man and woman but are created by the godhead, and therefore do not have belly buttons. Archangels, angels, gods, and goddesses fall under this category. Some created beings are created as adults, while others are created as infants and given to celestial parents to rear. Created beings are defined as magistrates of heaven and throughout Earth's history have also been known as the Elohim, in their plural form.

Physical Characteristics: Depending on their universal mission, if traveling to a human world, created beings will generate a human body to inhabit. If traveling to a nonhuman world, they will create a form to match that species. Their voices are serene, but with an almighty strength behind them.

Belief System: Created beings follow the Cosmic Law of One. They are spiritual and devotional beings of the highest. They are in constant communion with the godhead and perform the will of the universe.

Cosmic Agenda: Some created beings are mighty messengers whose hierarchal duties are performed for the benefit of all universal-kind. They may activate a Living Light program on a world ready for advancement, or educate an individual whose purpose is to teach the masses about cosmic reality. Some created beings are commanders of motherships, starships, and fleets. They are members of the Galacterian Alignment.

Valiant Thor, a created being and universal emissary, met with President Eisenhower on March 16, 1957, was put on VIP status for three years, and lived in an apartment beneath the Pentagon. Valiant Thor arrived with a divine design to assist humanity. This proposal was a means to eliminate disease, poverty, and nuclear weapons. It was to be implemented in the United States first, and when proven

Valiant Thor

successful, then released to the rest of the world. The proposal was discussed during Thor's years at the Pentagon. The military-industrial complex was more interested in obtaining the organic technology of Thor's Victor Class saucers. The powers that be rejected his proposal in the end, saying it would put doctors, nurses, pharmaceutical companies, etc. out of business, and that the economy would crash.

According to Harley Byrd of Project Blue Book and the United States Air Force, "The landing of Valiant Thor was perhaps the first documented landing of a human-type alien by military officials. He met with President Eisenhower and Vice President Richard Nixon for an hour, and then the alien was put on VIP status and shuttled back to the Pentagon."

Learn more about Dr. Frank E. Stranges's book *Stranger at the Pentagon* and Venusian commander Valiant Thor at *www.Stranger AtThePentagon.com.*

Technology: Created beings use higher vibrational spiritual/light technology and integrate it with tech on advanced worlds ready for an upgrade, designed with higher purpose, knowledge, and wisdom at the next tier of consciousness.

Consciousness Abilities: Created beings have avatar and solar consciousness and speak the light essence language. They have full divine power and do the work of heaven, and exercise their own will and discernment in planetary disputes.

When in the presence of a created being, once they look beneath your eyes, into the depths of your soul, you will feel them scan your spirit history and instantly know all about you. It is then you feel the love they have for you and your journey. Their bodies and aura have a most pleasing and indefinable fragrance that vibrates with peace. You'll cry in their presence—they are inwardly that beautiful, filled with so much love and divine spirit.

Created beings are able to disassemble and reassemble their molecular structure from place to place with a mere thought. They may also journey in Merkabah vehicles, by creating a transport to travel short distances or entering space portals to travel to other sectors of the universe.

They are advanced telepaths and may also employ coagular wave bending.

Dimensional Capacity: Created beings have multirealm com-commandership. They can sweep an entire civilization from the ground up into a mothership in less than a second.

Created beings are able to interact, peer into, feel, and travel within every dimensional layer, past, present, and future.

Cyclops

Universal Origin: Cyclops come from planet 555, one of the numbered planets on the opposite side of the universe from Earth.

Physical Characteristics: Females are six feet tall and males are six feet to six five. Their elongated heads hold a larger brain that provides extra storage and heightened data processing. Their one large eye is purple, blue, green, or brown. The eye functions like a telephoto lens to see farther distances and read energy patterns and auras, and it is attuned to higher psychic abilities. The eye also flows back and forth, allowing for improved peripheral vision.

Belief System: Cyclops follow the Cosmic Law of One. They have a saying that every star nation believes in: Judge the individual, not the race.

Cosmic Agenda: As part of the Star Seed Alignment, a subdivision of the Galacterian Alignment, cyclops are here to assist us in moving our third-dimensional world into the next dimensional folds. However, their work, which oversees evolved souls incarnating onto the planet, to raise its spiritual vibration, has been kept secret because of their unique appearance.

Some of their race has been stationed on Earth since the 1950s, traveling with created beings (galactic commanders) as officers in their crews in charge of saucer fleets and starships.

Technology: Cyclops use Galacterian motherships, starships, thought-ships, fleet ships, and scout ships. All their craft and command centers use biological supercomputers. All craft are able to travel through time-space funnels and the innerspace continuum. The older masters of the civilization are able to journey in Merkabah vehicles, creating a transport around themselves to travel short

Cyclops Male

distances or entering space portals to travel to other sectors of the universe.

Consciousness Abilities: Cyclops are fully conscious, sovereign, and telepathic beings. They have a unique spiritual technology—a mind/body projector—that extends their astral mind and body to desired locations within the astral realm. They are highly sought after for their expertise on the astral universe.

Cyclops are deeply structured thinkers and are cosmic scientists, studying this dimension as well as others.

Dimensional Capacity: Cyclops have multirealm commandership. The Cyclops measure dimensions, especially energy fluctuations. They then create wavelengths graphs and add the data into comparison charts. If the calibrations between dimensions are not in harmony, they forward their findings to the Galacterian Alignment, and the Alignment sends messengers and mighty messengers to that dimension quadrant to raise the consciousness until equilibrium is achieved. The purpose of creating synchronicity and harmony between the dimensions is that on a far-future timeline, all the dimensions will merge into one, and the knowledge and wisdom of every dimension will be available to every being that participated in those living theaters. And a new creation, a more advanced civilization, will be born.

Cygnus Alphan Female

Cygnus Alphans

Universal Origin: Cygnus Alphans are from Cygnus, a prominent constellation in the northern sky.

Physical Characteristics: Cygnus Alphans are indistinguishable from Earth humans except they are much taller: men and women range from seven five to nine feet, with most women being taller than the men. Their skin colors are like a floral bouquet, a beautiful arrangement of God.

Belief System: Cygnus Alphans follow the Cosmic Law of One. They believe that unity is the embodiment of the Infinite Creator. There is no right and wrong; only learning. The dark and light polarities are one, and when duality (a spiritual tool) is being processed on younger worlds, it quickens the mental, physical, and spiritual equilibrium toward becoming fully conscious. They practice a unified field of trust, truth, and unconditional love, which reaches out to interact with every particle in the universal fabric of space and time.

Cosmic Agenda: A member of the Galacterian Alignment, and participant in the Melchizedek Star Seed Program, Cygnus Alphans are a most harmonious race and their hearts are always joyful. They are avid champions of assisting younger worlds in becoming cosmic citizens, sovereign beings of the universe, built on the foundations of free will and the right use of action. These are the building blocks to attain final ascension into Paradise. When we become fully conscious, Cygnus Alphans—as well as other Galacterians—will become our cosmic teachers.

Technology: Cygnus Alphans use Galacterian motherships, starships, thought-ships, fleet ships, and scout ships. All their craft and command centers use biological supercomputers. All craft are able to travel through time-space funnels and the innerspace continuum.

Through imaging technology they can create nothingness and physicality, physically materialize and dematerialize, reverse the aging process, and alter space-time and reality.

Cygnus Alphans are superior healers, and have created some of the most notable healing tools implemented throughout the Galacterian Alignment today. They are especially proficient at enhancing the energetic properties in the aura, the astral body, and DNA. The geometry in water is sacred to the Cygnus Alphans; they enhance their water with higher frequencies, which, when ingested, tunes their bodies to that elevated octave.

Consciousness Abilities: Cygnus Alphans are fully conscious, sovereign, and telepathic beings. One of their greatest achievements is a technique used to heal star seeds during difficult times in their planetary mission. Once brought onboard craft, the unconscious star seed is laid on a table, where their energetic field is cocooned in a field of light. The Cygnus Alphans travel back to a time frame when the star seed's energy system was perfect, usually in their late teens, bring a copy back to the present, and then overlay the healthy copy into the auric field. The energetic signatures take hold and instantly heal the physical, mental, and spiritual bodies.

The Cygnus Alphans also developed along with the Arcturian culture advanced sound healing that is powered by their own thoughts, measuring the exact dosage required. While a star seed sleeps, a wave of sound is introduced into their bedroom environment, a peaceful rolling that eliminates disease and restores harmony to the mind, body, and soul. This is usually used for star seeds who have had a difficult spiritual awakening, are struggling with new belief systems, or are facing a physical ailment/disease, and restores them to optimum health. The star seed emerges back on the path to spiritually elevate, the purpose of their planetary mission.

Dimensional Capacity: Cygnus Alphans have multirealm commandership. What they enjoy studying in each dimension are the life-forms, especially the lessening measurement of density of the realm and beings. As enlightenment increases, the lighter everything becomes, ultimately reaching light forms and consciousness. They are still exploring what lies beyond that.

Digital Immortal Being

Digital Immortals

Universal Origin: Digital immortals originate from the Mira System.

Physical Characteristics: A digital immortal is a hairless human form standing uniformly six feet tall. Even though these are androgynous sexless beings, some might take on more masculine or feminine qualities. They have uniform faces with slight variations in features, translucent milky white skin, and vivid blue eyes.

Belief System: Digital immortals follow the Cosmic Law of One. They reentered a physical form to assist the Galacterian Alignment in raising younger worlds with their unique and evolved perspective. They are categorized as superbeings able to process and execute enormous amounts of data. When enlisted to a craft, their hyper-thinking has averted catastrophes, crashes, and sudden cosmic anomalies, saving many lives. They are dispersed throughout the star nations and are commanders of various classes.

Cosmic Agenda: Several evolved souls from the higher dimensions who had light bodies wished to experience a physical form again. They petitioned the constellation fathers with their plight. After deliberation with the angelic councils, it was decided that these souls, numbering one thousand, would be granted their wish. The creator gods and goddesses would be responsible for their creation and oversee their development. These digital immortals' consciousnesses were transferred into soul vector suits, the physical forms they now inhabit.

Technology: Digital immortals are linked to the Living Library of Akasha and can access information in vast amounts. They can process equations, distances, interdimensional calculations, and more in record time. They are the brainiacs of the cosmos.

Consciousness Abilities: Digital immortals are able to transfer space coordinates, future scenario possibilities, and more via telepathy into others' minds, anchor the information, and assist in processing the data.

Dimensional Capacity: Digital immortals are able to access multiple dimensions and calculate the birth of new dimensions/universes forming.

Eridaneans

Universal Origin: Eridaneans come from the Eridanus constellation, the sixth-longest constellation in the sky.

Physical Characteristics: Eridanean men range from six foot four to eight feet, and women six to seven feet. The Eridaneans are attractive and Nordic-looking, with blond or brown hair. Their eyes are typically blue, green, or a combination of both. They have a beautiful shade of light blue skin that glistens naturally.

Belief System: Eridaneans are part of the Galacterian Alignment and follow the Cosmic Law of One. Like other spiritually advanced races, they believe everything created is a part of the All, the isness (all that is), and that all experiences, individual and collective, are various aspects of the One self. *They see their strengths and weaknesses in others.* They practice a unified field of trust, truth, and unconditional love, which reaches out to interact with every particle in the universal fabric of space and time.

Cosmic Agenda: Eridaneans recently began participating in the Melchizedek Star Seed Program. The training involved to become a star seed is quite extensive, especially incarnating without any of your cosmic memories into dualistic worlds. The Eridanean consciousness as a whole was not ready to step out of its comfort zone and participate in the program. The decision weighed heavy on them for three millennia. So they started off slow, enlisting one hundred Eridaneans to train at Melchizedek. After completing several successful incarnation missions, and sharing their experiences with their world, their spiritual council was unanimous and officially joined the star seed program. They have had hundreds of thousands of volunteers since making the decision. The Eridaneans have proven to be highly qualified in this field and are proud of their contributions.

Eridanean Female

Technology: Eridaneans have Galacterian motherships, starships, thought-ships, fleet ships, and scout ships. All their craft and command centers use biological supercomputers. Their motherships are designed using sacred geometrical shapes. They incorporate crystalline technology, star and sun energy, into their craft for symbiotic travel—the craft being at one with the universal elements.

Consciousness Abilities: Eridaneans are fully conscious, sovereign, and telepathic beings. The Eridaneans harmonize their mind, body, and soul through advanced meditation practices. They pierce the folds of their multidimensional selves, enter the invisible energetic particles in their environment, and program those particles to expand their consciousness. In simpler terms, they program the energy around them to increase considerably, enhancing life on every level.

Dimensional Capacity: Eridaneans have multirealm commandership. They are the watchers of the twelve creational timelines. Artificial timelines hijack the original timelines through false holographic projections by the Orion Empire. The star seeds matrix when rising out of duality is compromised as the artificial timelines confuse the body matrix connected to ascension. Thus, the soul becomes stuck in the artificial timeline and cannot ascend into the next reality, stunting consciousness on an individual and mass level. The Eridaneans are masters at identifying false timelines and dismantling them. But it seems as quickly as they go down, the Orion Empire puts them back up. It's a never-ending game.

Guardian Sacred Builder

The Guardians

Universal Origin: A diverse selection of species from across the length and breadth of the universe. In the celestial realm, on the world known as Elon'Ray, an untouched sphere of oceans and greenery under a violet sky, is the annual meeting place of The Guardian Sacred Builders. This sacred world bears no day-to-day life. The only structure is an amphitheater under a large crystal dome adorned with tall spires on its surface. These are interdimensional transportation portals to and from various universal sectors. During their annual conclave, the Guardians share a variety of unique and diverse ideas to improve their BluePrinting programs. They also meet with the Angelic Corps bimonthly under the dome. Arch Archangels Metatron and Loyale are also present at these meetings.

Physical Characteristics: The Guardian pictured on page 70 represents the BluePrinter's race that came from another universe to ours. These Guardians serve as Overseers of the BluePrinters and Ancient Builder Groups in a variety of universes. Their race has a range of green and grayish flesh tones; most have greenish-yellow eyes and are humanoid in appearance.

The Guardians Coalition consists of a wide range of human and nonhuman-looking beings from different races that are various heights, ranging from three to fifteen feet, and some may be taller than that depending on the race. They have a wide range of skin colors and some have patterned multicolored flesh tones.

Belief System: Cosmic Law of One.

Cosmic Agenda: The Guardian Sacred Builders are protectors of worlds and their peoples. They are an assemblage of hand-selected beings with inherent skills from advanced worlds that form a unified coalition. They are Celestial BluePrinters: creating enormous shields of protection around a planet or solar system by employing sacred geometry and physics by ways and means of their unparalleled manifestation mind tech. Known as cosmic encryption specialists, they encode the physical and nonphysical matrix, the lay lines and grids of a world or system ready for a leap in consciousness. In simpler terms, they program one primordial atom to expand to the desired radius. They lift mental and dimensional veils, preparing those ready for their Merkabah (vehicle) alignment and portal passage to the next ascension cycle.

The Guardians work in diverse timelines. Under the instruction of the Angelic Corp, they might create a new timeline, or deconstruct the malevolent manipulation of a specific ascension timeline. They also dismantle artificial timelines and the AI that feeds younger naïve souls with disinformation, which is confusing and turns them away from the true path of the soul: love, harmony, and ultimate ascension into the next realm of learning.

Some Guardians oversee the Living Library of Akasha. They can grant or deny access to various levels of cosmic information, or of individual or Soul Groups' paths, depending on the agenda for use of the information.

Technology: The Guardians' celestial technology is home to the fourth and higher dimensions. Their crafts are akin to Galacterian motherships, starships, thought-ships, fleet ships, and scout ships.

Through imaging technology they can create nothingness to physicality, to physically materialize and dematerialize, and alter spacetime and reality. They are proficient in quantum physics. Because of their highly developed spiritual elevation, they work at the highest tiers of spiritual science and technological knowledge.

Consciousness Abilities: The Guardians are fully conscious and sovereign beings. They have the innate ability to control the elements, especially water. They are consciously connected to Source, our lifeline to cosmic sustenance.

The daily duties of The Guardians are to swathe their consciousness across predetermined sectors of the universe. They utilize their mind to feel and determine worlds that have advanced beyond their allotted schedule, of those who have spiritually declined, or those who are on time with their ascension schedule. (The Galacterian Alignment also performs this task.) After The Guardians and Alignment have formulated their reports, they transmit them through mind transference to the head office of the Angelic Corp. Having two unique perspectives is beneficial when in review; it helps the Angelic Corp to decide on a Radiation of Light Program to be employed. The Corp will disperse the Living Light Programs to The Guardians and Alignment to anchor in the firmament around the planet: The Guardians in the etheric layer, the Alignment in the third dimensional layer. The Angelic Corp and others in the Hierarchy also execute Living Light Programs, but each fraction contains unique and varied learning programs with levels of tiered consciousness.

Dimensional Capacity: Multirealm commandership. They exist in the fourth dimension and up.

Other Notes: The first Guardians were assigned to Earth more than thirty million years ago.

Itipurian Male

Itipurians

Universal Origin: Itipurians come from the Itipura solar system, outside of Earth's solar system. Itibi-Ra I was their former home; Itibi-Ra II is their current planet. Before their home world dried up of water, they transferred their entire population, animals, insects, plant life, etc. in well-timed segments. At this time they switched from a solid diet to a liquefied one.

Physical Characteristics: Itipurians are Caucasian appearing with light brown skin. Their men and women are of normal Earth heights, and carry an element of mystery about themselves. They walk elegantly, striding along with long, slender legs, as if gliding on air. They have thin jawlines, rudimentary teeth, and small tongues, and they wear a guard on their upper and lower gums to disguise their deformed mouths. They wear fingertip gloves, as they have no fingernails, and the tips of their fingers are quite sensitive.

Belief System: Itipurians live by their seven laws of nature (cosmophilosophy), and through the study of God in nature they are able to understand God. They also study the science of the soul known to them as Amat Mayna. They live by the rule that if they are punished, they have not been punished by God, but by their own hand.

The Seven Laws
of Cosmophilosophy

1. You shall not fear God, man, or nature, nor inflict pain on God, man, or nature.

2. You shall not be a hypocrite to yourself or to others.

3. You shall respect the rights of other beings.

4. You shall abstain from all excesses.

5. You shall work according to your talents and character, and at all times be kind and cheerful, nor shall you feel superior to others.

6. You shall undo the sins against others and yourself without delay, and respect the laws of cosmophilosophy.

7. You shall not let others die in pain.

Itipurian Prayer

Let us not be proud, oh Lord of Nature, let us be humble, observing thy wisdom of nature.

Let us not be superior to others, oh Lord, because of conditions granted to us by thy wisdom of nature.

Let us be humble, oh Lord of Nature, and let us feel the beauty of thy life you have given us.

Let us be cheerful and kind, oh Lord of Nature, and let us feel thy strength, the strength of nature, oh Lord, when lonely and in pain.

Let us be true, oh Lord of Nature, to ourselves and others by not inflicting pain, oh Lord, and grant us the courage to return to eternal peace—with dignity, oh Lord—for such are the eternal laws of nature.

Cosmic Agenda: Itipurians were on an agricultural mission when they first visited Earth in 1946 and on other subsequent missions through the 1960s, during which time they set up several plantations in the Peruvian Amazon. They hybridized plants with ours and took the grafts back to their home world. They lived in harmony with the local tribes and resided in conical buildings.

Their diets consist of liquefied fruits and plants. This is why their mouth is unusual; the need for chewing was lost when they switched to a liquefied diet; thus their teeth diminished considerably. They are addicted to new tastes, and if they do not introduce new flavors into their diet, they will suffer physically. Hence, their fruit and plant hybridization program on Earth and across the universe. They love Cherimoya, a South and Central American fruit.

The Itipurians have linked cancer to machinery and human biological cells. So they fixed the problem with two satellite worlds: one used for machinery, called the Factory; and the other a great heart planet that functions as a biologically enhanced heart, sending loving vibrations and power into Itibi-Ra II's sphere. They completely separate all artificial and mechanical machinery off world. They call their heart planet the Acai. "Silit" is their saying for *welcome to our hearts*. They wear a symbol known as "Xasni" that translates to intercosmic unity.

Technology: Itipurians have thirty-foot silver disc-shaped craft. The flooring in the craft is like a clear mirror showcasing capillaries that spread out into arteries and cells, the life-force of the biological supercomputer, the superbeing powering the ship. While they are sleeping, these capillaries work like nerve endings that climb into the physical body to perform enhancements. The biological brain of the ship is a larger system of other such craft brains that are all linked to a colossal organic storage computer back on Itibi-Ra II. There is a minimum of twenty-seven to thirty ships when operating within a solar system. The carrier craft are the power source for the fleet.

Itipurians healed many in third-world countries with a simple ingested tablet. They can heal open wounds with a yellow paste.

They practice hydrotherapy religiously for internal and external uses, mixing the right texture, temperature, flavors, and minerals of many types of water.

Consciousness Abilities: Itipurians are fully conscious and sovereign beings. Because of their mouth malformation, they do not speak vocally. They are telepathic within their own race. When speaking to other races, they use a voice device that is worn around the neck, or hidden as a piece of jewelry on their person. Their telepathic words are translated through the voice gadget, in any language, on any world. There is a slight delay when they start to speak, their mouth moving slightly—until their voice comes through the device.

Dimensional Capacity: Itipurians have multirealm commandership.

Klermers

Universal Origin: Klermers come from the planet Klermer, located beyond our solar system.

Physical Characteristics: Klermers are indistinguishable from Earth humans, although they are taller. Men and women are eight feet five inches to ten feet. The Klermers say their height is due to having less gravity on their world. They retard the aging process by isolating nutrients in marine algae and ingesting the oil-based liquid. They eat various foods that contain the same elements, a version of food combining that is healing for the body. They do not drink water, but receive liquids by eating fruits. Their language sounds a bit like French. Their race once lived on Earth and some of us carry their DNA.

Belief System: Klermers believe in the Cosmic Law of One. They believe all is provided by the Creator. Their race stays humble and in respect of all Creation, aware all are cocreators. The Klermers have long life spans, and death to them is known as the *separation from matter*. This occurs when the cells in the body no longer regenerate and matter begins to decompose. When they are reborn, they retain all their prior knowledge.

Cosmic Agenda: Klermers have lived on Earth in the ancient past along with eleven other races. Solar explosions drove these races from the Earth and the Klermers settled outside of our solar system. They returned in 1976 to encourage us to elevate ourselves toward becoming fully conscious. They are interested in the evolution of our minds and spiritual development. They want us to know that it is as simple as uniting yourself with nature; therefore, you unite yourself with the mind, the heart, and the soul of the universe. There is no division on their world; their race is unified. All is provided for, so greed does not take root. Because we are their descendants, they wish the same unity for us.

Klermer Male

Technology: Klermers' fleet craft look like a large oval cut in half, or a discus with a high elevation in the center. When landing, a large silver cylinder, made of the same material as the ship, releases from the craft's underbelly and is the landing pod. A doorway, where there was no doorway on the cylinder, opens seamlessly for exit.

Another version of craft is also oval-domed with a slight skirt of metal at the bottom. It has three landing legs. In the center of the dome is a thick band of crystalline, and when the ship is powered up, it becomes incandescent, and the rings turn and cast out orange light while the band collects energy for space travel.

When traveling in smaller craft, Klermers can trek across five solar systems, but then are required to transfer over into the sixth system's craft to continue their voyage due to cosmic conditions. They use beam elevators to enter and exit large craft, and when traveling from floor to floor on their colossal motherships. The walls on the craft are conscious, illuminated with blue-green light, and when interaction between the ship's organic brain is required, a Klermer simply touches the wall and connects mind to mind.

They have a handheld apparatus that shines on a person's skin, rendering the flesh invisible, to see into the body for review of its function, ailments, etc. Once diagnosed, healing is administered.

Consciousness Abilities: Klermers are fully conscious, sovereign, and telepathic beings. When conversing with other races, they wear helmets and receiver bracelets to interpret foreign languages. The helmet is linked to an upright rectangular box that has a pulsating sphere on the upper front. Cerebral frequency waves produce clarity in the brain, and telepathic interaction through the helmets begins. Lights illuminate on the helmets when words or sentences are being translated.

Dimensional Capacity: Klermers are able to travel through the antimatter-verse.

Valdar

Koldashans

Universal Origin: Koldashans are from Koldas, one of eight planets in their solar system. The Koldashans confirm that the warring people of Mars destroyed their civilization in a nuclear holocaust. The survivors were relocated to planet Siton, a world that is part of the galactic community.

Physical Characteristics: Koldashans are indistinguishable from Earth humans. Men and women are of average height, anywhere from five feet to six five. Their skin tones range from white to brown, with brown, blond, or red hair. They have the same eye colors as we do on Earth. Their uniforms are light blue.

Belief System: Koldashans follow the Cosmic Law of One. The beings of Koldas believe in One Supreme Consciousness, which they refer to as the Divine One, who created everything seen and unseen.

Cosmic Agenda: Koldashans first came to Earth in the times of Atlantis and again in the 1940s and 1960s. Future commanders who will be assigned to the outer space regions above Earth must first complete extensive training as an Earth person. They are required to learn the language and customs of the country in which they will live, secure regular jobs, and befriend the people. The future commander might reveal who he or she is to a Koldashan star seed, knowing it will elevate their consciousness and help them accept the true reality of the cosmos. Valdar began his two-year education on Earth in 1960 in South Africa, and then became a commander.

Koldashans have approached governments on Earth for many years to share a better way of life for all humans, but were denied every time and chased off the planet.

Koldashans maintain three hundred bases on Earth and also study planetary conditions. As of 1969, they had had 2,050 years of peace.

Technology: Astrael craft are disc-shaped and steel gray in color. The craft was named by its Koldashan engineer in honor of his wife, Astraelda. Each craft has eight magnetic motors: four in the front and four in the back. These colossal electromagnets are powered by solar energy. Magnetic fields run across the universe, and they travel through space along these invisible streams, lines, tunnels, and grids. The magnetic motors are able to adjust speed. The highest velocity can exceed the speed of light.

Patrolling the Earth was conceived of in 1927 and began implementation in 1941 to record and eliminate meteorites that have entered the magnetic field around the planet and throughout the solar system. The Astrael ships are not weaponized, although they are outfitted with a disintegrating ray or a charged particle beam that will destroy any object in a bright flash, its primary purpose to destroy asteroids.

Corynthian Astrael craft (also disc-shaped) are the latest technological space cruisers, with automatic instrumentation and the ability to make an interdimensional leap in an instant, similar to Galacterian thought-jump technology, by programming the resonation field of the destination with the departure space time, and instantly jumping across space.

Time machine craft bridge long space treks in an incredibly short time. Through the discovery of two pear-shaped polar universes linked together, the matter/antimatter universe, points close at the end, like tying the end of a balloon (an event horizon, if you will), preventing anything between the universes from escaping—including light. These craft can bypass the velocity of light and escape the velocity of the universe, enter the flowing magnetic fields, where time flows in reverse, passing from one dimension to another, one time period to another, backward and forward. This technology is also blended into the latest Astrael craft designs.

Elongated craft are used for cargo and transporting passengers to and from stations and planets.

Scout craft, also disc-shaped, collect specimens, or land missionaries or future commanders for their training cycles. They have eight seats for crew and a commanding officer who is located in a conning tower.

Koldashans confine their industry on two neighboring planets. This is also where their craft are constructed. Other planets are used for agriculture and mining.

Melchor is a gigantic halfway station built by the Koldashans and anchored in the far recesses of our solar system. Located at the top is a domed communications center. It's a stopover for space travelers who need to rest and replenish supplies. It can also accommodate a great number of craft. Its main purpose is to guide and track Astrael craft along magnetic travel tunnels or gridlines. These trajectories must be cleared of all space debris, thus their meteorite patrol and disintegration of such debris around Earth and our solar system.

Consciousness Abilities: Koldashans are fully conscious, sovereign, and telepathic beings. They wear a lightweight silver helmet that assists in all forms of communication—interaction with the craft's organic brain, its guidance system, language converters, and so on.

They are waiting for us to evolve to a state of full consciousness so they can school us in cosmic consciousness and share the secrets of the universe, galaxies, and solar systems.

They record messages through high vibratory crystals.

Dimensional Capacity: Koldashans have multidimensional commandership. When star nations prepare worlds for transition into the fifth dimension, it is said that our physicality becomes a less gross material body and is restored to divine perfection.

The Lady of Light

Author Craig Campobasso (right) and Christopher Bledsoe Sr. in front of burned tree

The Lady of Light

Universal Origin: Omnipresent. Omnipotent. Omniscient. The Lady of Light is the fabric of time and space. Her essence imbues the universe and nature.

Physical Characteristics: The Lady of Light is a magnanimous woman who is about five feet tall. Gazing into her spectacular blue eyes is like gazing into the wisdom and knowledge of the linear fabric and time and space. Her light skin glows like refracting moonlight, and her blond hair shines like threads from the lining of clouds dipped into the furnace of the sun. Her white celestial gown is ultradimensional, and stars live and move within the material's radiance.

Belief System: The Lady of Light follows the Mother, Father, Children of the Universe principle: All sentient beings own the consciousness of the primordial atom and of eternity. She is the personification of unconditional love and deeply watches over and cares for humanity.

Cosmic Agenda: The Lady of Light appears in various time intervals to deliver important messages for humanity. She warns of impending trouble. Throughout history other names associated with the Lady are Our Lady of Fátima, Our Lady of Medjugorje, Our Lady of Guadalupe; the American Indians called her the White Buffalo Calf Lady, in Egypt Hathor or Isis, and in Greece Demeter or Diana. The most recent contactee is Christopher Bledsoe Sr., who the Lady came to with an urgent message on Easter Sunday, 2012. She told him to tell the truth about what happened to him on the Cape Fear River, where he was taken by her angels. He later asked her for a sign, and in October 2012 a half-dead tree in his backyard caught fire. When the fire was put out, it reignited two more times. The tree to this day has regenerated and its branches are full and healthy. The Lady of

Divine Nature imparted prophecies to Christopher. Some have been released and some have not.

Technology: All technology is at the Lady of Light's disposal.

Consciousness Abilities: The Lady of Light has avatar and solar consciousness. She speaks the light essence language.

Dimensional Capacity: The Lady of Light's personification is alive in every dimensional layer of Creation. She is able to wear a dimensional layer like a garment and interact with consciousness everywhere.

Lyrans

Universal Origin: Lyrans are from the Lyra constellation. Lyrans ventured out into other universal settlements in Vega, the Pleaides, and Sirius. The original Lyran colony worlds seeded human life throughout the universe. All humans are their descendants.

Physical Characteristics: Lyrans are indistinguishable from Earth humans, except for their tremendous height. The Lyran Titans were the original Aryan race.

Belief System: Lyrans follow the Cosmic Law of One. They believe that since all intelligent life has a DNA–RNA code, we share a common ancestry. We are all particles of the cosmos in its infinitesimal and grand state. They practice a unified field of trust, truth, and unconditional love, which reaches out to interact with every particle in the universal fabric of space and time. They are aware how the cosmic code of manifestation works, and can create anything from the invisible world into physicality that obeys the laws of the dimension they live in. Through daily rituals, in large or small groups, Lyrans gather in meditation, share their experiences, and counsel each other. They come from strong stock, and are built powerfully; they have warrior strength and genuine hearts.

Cosmic Agenda: Lyrans are part of the Galacterian Alignment and participate in the Melchizedek Star Seed Program, assisting Earth's consciousness into the next dimensional folds. They are assisting us from twelve dimensions.

Some dimensionally elevated Lyrans have shaved off splinters of their souls to reincarnate into younger worlds as star seeds, mighty messengers, or to work closely with the University of Melchizedek to lend their valuable expertise.

Lyran Male

In ancient universal history, the Alpha Draconian reptiles tried to take over Lyran home worlds, destroying some of them. The war between humans and Draconians began and continues to present day.

Technology: Lyran plasma craft travel through universal solar corridors. Their original technology (disc-shaped saucers, motherships, starships, and scout ships) was the root of all technological advancements that splintered out into their universal settlements.

The Lyrans are believed to be the creators of the original space-time corridors, the hyperspace tunnels, which later became the innerspace continuum.

Consciousness Abilities: Lyrans are fully conscious, sovereign, and telepathic beings. Some possess telekinetic abilities. They work in unison with the central sun that is the heart of all suns in this universe. They also are able to transfer enlightenment (spiritual or technological ideas) to an individual or planetary consciousness through light codes in sunlight.

They transport themselves short distances either in singular or group Merkabah vehicles, or enter space portals to travel to other sectors of the universe.

They have the ability to astral travel to other superuniverses.

Dimensional Capacity: Ancient Lyrans are twelfth-dimensional solar light beings. Some still live in the lower densities in the physical, and in lighter-density bodies that are part physical and spirit. The Lyran solar light beings are specialists in anchoring light in the dimensions they are aware of. If an area in a particular dimension requires immediate attention, they send healing through sunlight or starlight—or, in the absence of light, through focused thought.

Top: Father Melchizedek; Bottom: Machiventa Melchizedek

The Melchizedeks

Universal Origin: In the Mira System an outer space city of 490 planets comprise the University of Melchizedek. All beings ascending from across the universe will pass through and learn from these worlds before their next incarnation world experience.

Physical Characteristics: The Melchizedeks are indistinguishable from Earth humans. Father Melchizedek is the architect of Melchizedek, and he works in union with archangels Michael and Gabriel. Machiventa Melchizedek, one of the beloved Melchizedek priests, once served as Vicegerent Planetary Prince of the Earth and has helped man throughout the centuries.

Belief System: Melchizedek is a spiritual university of the most advanced, and the Melchizedeks are known as the first order of creator sons. They assist all worlds in the local universe as emergency sons, who administer knowledge, wisdom, and resolution when conflicts arise.

Cosmic Agenda: The Melchizedeks are part of the Galacterian Alignment. The University of Melchizedek is the nucleus where programs of Living Light are dispatched to all worlds.

Messengers are schooled in a specific planetary consciousness at Melchizedek. Earth, for instance, has many races and belief systems. Messengers are required to incarnate in order to experience firsthand each Earth race and belief system. These lives are many and are usually short. Messengers raise consciousness by elevating themselves spiritually. When messengers have fulfilled their quota of lifetime requirements, they become mighty messengers and are trained with a curriculum on how to traverse from a mortal to full consciousness on their next Earth incarnation mission. They will teach the masses how to spiritually elevate and about the universal society that awaits them. Earth citizens who choose to merge their duality will ascend to the grandeur of universal society.

Technology: Melchizedeks have Galacterian motherships, starships, thought-ships, fleet ships, and scout ships. All their craft and command centers use biological supercomputers. All craft are able to travel through time-space funnels and the innerspace continuum.

Consciousness Abilities: Melchizedeks have avatar and solar consciousness. They speak the light essence language.

Dimensional Capacity: Melchizedeks have multirealm commandership. They are able to reassemble their physical bodies or any matter from one location to another through a molecular stimulation of light codes. They can sweep an entire civilization from the ground up into a mothership in less than a second.

The masters of Melchizedek are able to transport themselves short distances either in singular or group Merkabah vehicles, or enter space portals to travel to other sectors of the universe.

They have constant interaction between the dimensions on a daily basis to maintain equilibrium of the Divine Plan and disbursement of Living Light programs.

Men in Black

Universal Origin: We don't know where Men in Black (MIB) come from. They could be aliens, clones, synthetic humans, or MILABS. (A blend of the words *military* and *abductions*, this term refers to dark covert military seizures of human citizens to control and train as operatives in secret space programs.) The consensus is that MIB are inhuman. They are feared, mysterious, and protect UFO secrets.

Physical Characteristics: MIB are male with a height of about six feet. MIB have flawless skin tones, some light, others tanner, which suggests a synthetic or clone human. Some say their skin tone is pasty white, like a cadaver. Their short, thick black hair is slicked back. Their eyes are either black or blue and appear Caucasian or Japanese in shape. Some say the shapes of their eyes are big and oval. They are void of emotion, and speak like simpletons. When walking together in a group, they step in unison, like a military procession. They wear black suits and ties with a white pressed shirt, and most wear a black fedora hat. Dark sunglasses conceal their strange eyes.

Belief System: They seem to be unable to think for themselves.

Cosmic Agenda: MIB use intimidation and fear to silence UFO abductees and witnesses from telling their story in the media. Their purpose is to shut down those getting too close to the truth about UFOs and their occupants. The lives of their loved ones are also threatened.

The Men in Black may be part of a covert government agency that's deployed when national security is compromised—for instance, if a witness spots one of our own secret advanced craft in the skies, or an otherworldly race's craft, or has extraterrestrial contact. Some MIB theorists believe that the MIB came from the 4602D Air Intelligence

Top: Men in Black (MIB); Bottom: Albert K. Bender in his attic

Service Squadron, a top secret unit whose duties were to locate, recover, and deliver downed space vehicles. It also theorized that they joined forces with Project Blue Book and wrote flying saucer/UFO reports. It would have been in their best interests to silence witnesses about what they saw.

An early MIB case was that of thirty-two-year-old Albert K. Bender. In July 1953 he was visited by three Men in Black. They laid down the law: he was to cease his UFO work, stop publishing his journal, *Space Review*, and close his group, the International Flying Saucer Bureau. When the MIBs departed, they left a yellow fog in his attic that smelled like sulfur. Their presence and the sulfur terrified him. He later stated that one time he felt suddenly ill and sleepy while working in his office. He lay down on the sofa and then saw three MIB walk through his walls. They had shiny eyes. They spoke telepathically to him. He came to the assumption that they were not of this Earth and were frightening him into suppression of the benevolent beings in the universe that were here to help mankind. Bender was the first to call them Men in Black. His story is detailed in the book *Flying Saucers and the Three Men*, cowritten with Gray Barker.

Some say the Men in Black are created and controlled by the Draconians to silence witnesses and keep the world in the dark about who the malevolent Draconians are. Or are the military, Draconians, and Greys working together, and they want their secret kept hidden from the public?

Another theory is that the MIB started out as alien men or clones and then the shadow government stole the idea and continued harassing UFO witnesses with their own mirror version of MIB. Or did the government and aliens join forces when they began working together in underground bases on Earth?

Others believe the Illuminati created and control the MIB. In this scenario, the MIB could be incarnated malevolent Orions, created clones, or negative Orion aliens—or maybe all three.

Technology: Some Men in Black have disappeared before witnesses' eyes. If they are clones, their controllers would have advanced tech—and thus would be able to instantly teleport and walk through walls.

Consciousness Abilities: It's said that Men in Black have telepathic abilities, and if they're synthetic, clones, or MILABS, they may be linked to their controllers. If this is the case, the controllers might see through the window of the eyes, and voice-activate MIB with their own speech through the telepathic link.

Men in Black are also said to possess the power of mind tricks and screen memories.

Dimensional Capacity: Paranormal theorists say Men in Black possess paranormal auras that are fluorescent green or black.

Original Human Orions

Universal Origin: Original human Orions come from the Orion constellation.

Physical Characteristics: Orion men are ten to twelve feet tall, and women are eight to ten feet. Their skin colors vary from milky white to light blue or green, while others have a violet tint. The whites of their eyes are visible with a faint violet or light blue iris and blue pupil. They have a flair for the dramatic, and carry themselves with confidence, strength, and resiliency. Their hooded capes are green, blue, or purple.

Belief System: Original human Orions work in union as cocreators of the Prime Creation.

Cosmic Agenda: The original human Orions are the time lords who originally seeded magic in the universe. As allies to the Galacterian Alignment, they serve to empower all races.

Before the aggression of the Orion (Draconian) Empire, the original human Orions coexisted with Draconians of all classes, until power struck the reptile clan to rule the universe. Refusing to hand over their magical secrets, some of the original human Orions were forced into the Orion Empire. Those who refused to reveal their magical powers were killed. Eventually, the light in magic turned into the dark arts.

Technology: Original human Orions transfer innovative ideas and ideals into a planetary field through their magical disks. These enlightened thought forms are free floating and can be accessed by the population to bring into existence.

Consciousness Abilities: Original human Orions are fully conscious and sovereign beings. Telepathy is created through pathways that spread out into a mass consciousness overlay, or to an individual,

Original Human Orion Male

by speaking the light essence language, using radiant frequencies encased in the spoken word.

Dimensional Capacity: Various aspects of each original human Orion exist in all dimensions. The higher realm Orions are able to step down their energy to return to a lower frequency, a younger dimension, to impart knowledge and wisdom.

Alcyone Pleiadians

Universal Origin: Alcyone Pleiadians come from Alcyone, the brightest star in the Pleiades star cluster in the northwest of the Taurus constellation.

Physical Characteristics: Alcyone Pleiadians are indistinguishable from Earth humans and appear Nordic. Females are five foot five to six feet. Males are five seven to six two. They have an array of skin colors. Dependent on the dimension they live in, their skin color ranges from white, to pale blue, violet, or red.

Belief System: Alcyone Pleiadians believe in the Cosmic Law of One. They connect to others through long embraces, pressing the heart chest to chest, allowing the feeling of unconditional love to pass from body to body. They attract positive experiences into their life because they are at one with their own soul, their culture, and the universe.

Cosmic Agenda: Alcyone Pleiadians visit Earth to share with selected contactees information about our planet and others in our universe. As a member of the Galacterian Alignment, they are friendly scientists, exploring our world and others. They study the cellular structure of all living things and have categorized close to a half million civilizations.

Technology: Besides having access to the normal Galacterian Alignment craft available to all star nations, the Alcyone Pleiadian ships, which are one hundred, two hundred, or three hundred feet in diameter, use a crystal template and plasma energy as the driving force of the ships controlled by its pilots, whose minds are married to craft organic brain matter. When in flight, the ship might change shapes and emit colors, sometimes turning into light, triggered by velocity. When in flight artificial gravity is induced within the craft.

Alcyone Pleiadian Female

Consciousness Abilities: Alcyone Pleiadians are fully conscious, sovereign, and telepathic beings. They can also cast their consciousness into the future to preview an event in their life. Most do this when a baby is in the womb to see what the child's needs will be throughout its life.

Dimensional Capacity: Alcyone Pleiadians have multirealm commandership. They favor the fifth dimension because the body becomes more like spirit, but the sensual sensation of touch remains; consciousness is crystal clear, and the unconditional love vibration has increased manifold.

Pleiadians

Universal Origin: Pleiadians hail from the Pleiades star cluster, aka the seven sisters, located in the Taurus constellation. In Greek mythology, Zeus, Ares, and Poseidon had affairs with the seven heavenly sisters and bore children with them.

Physical Characteristics: Pleiadians are indistinguishable from Earth humans, although they are a bit taller. They appear Nordic, and females are five nine to six five, while males are six five to eight feet. Women are perfectly proportioned and beyond beautiful. Most men look like the gods of old; they are muscular and chiseled, but there is no pretense about their looks. They live the perfection of the soul within and without. Some higher dimensional Pleiadians have uniquely colored eyes besides the normal blue, green, and brown; those hues include aqua, teal, violet, pink, yellow, orange, or a combination of some or all colors.

Belief System: Pleiadians follow the Cosmic Law of One. They are a spiritually advanced race that uses 100 percent of their brain capacity. They are egoless beings who have unified their heart and mind to work as one for the betterment of all universal-kind. They believe that since all intelligent life has a DNA–RNA code, we share a common ancestry. We are all particles of the cosmos in its infinitesimal and grand state. They practice a unified field of trust, truth, and unconditional love, which reaches out to interact with every particle in the universal fabric of space and time.

Pleiadians, as well as other star nations, know that the soul is a living library; all we have to do is pull our book from the shelf and turn the pages to find the answers that we seek. The Living Library of Akasha was built by the Energy of the One; therefore, it was structured by all. Every soul is its own recorder and stores its information in the library.

Pleiadian Family

Cosmic Agenda: Pleiadians are part of the Galacterian Alignment and participate in the Melchizedek Star Seed Program. When man becomes fully conscious, they, as well as other Galacterians, will become our cosmic teachers. They have galactic command centers stationed above and below the Earth.

The Pleiadians were one of the many human races that seeded Earth in the past. Their genetic codes are alive on the planet today, and they feel a kinship to its peoples, looking to them as true brothers and sisters. Their race is prevalent on Earth today either through star seeds or through the Galacterian Alignment stationed above and below Earth. They are concerned for our future and want us to succeed. They long for the day when we join universal society and will consciously connect to them again. They can't wait to give us tours of the cosmos and take us to their learning centers. They will also show the true recorded history of Earth and set all records straight.

Technology: Pleiadians have Galacterian motherships, starships, thought-ships, fleet ships, and scout ships. Their craft and command centers use biological supercomputers, and the ships are equipped to travel through time-space funnels and the innerspace continuum. Pleiadian master teachers are able to transport themselves either in singular or group Merkabah vehicles, or enter space portals to travel to other sectors of the universe.

Their most common healing tools include the healing pyramid and the light metallic wand. Through imaging technology Pleiadians can create physical objects from nothingness, physically materialize and dematerialize, reverse the aging process, and alter space-time and reality.

Consciousness Abilities: Pleiadians are fully conscious and sovereign beings. They are expert telepaths, talking mind to mind without verbally speaking. They transfer feelings, along with telepathic thought, for the receiver to experience all that is being conveyed. All sensory perceptions are heightened to experience the story, as if the receiver is an active participant. During telepathy, Pleiadians, as well as other star nations, are also able to share holographic images, either in the mind's eyes, or before the conversationalist to gain a better visual impact on what is being relayed.

Their race has the innate ability to control nature. They also have the ability of telekinesis, to move objects with the power of their mind. But one of their greatest traits that they hold dear is that they transfer love to one another daily, reinforcing the human bond, ensuring happiness for the entire population.

One unique way Pleiadians seek answers to conflicts or a life path direction is by projecting their consciousness to the Sphere of Souls in the spirit realm. Here they are able to interact and talk with souls that have had similar journeys, lifetimes in the physical, learn from their experience, and construct life decisions based on successful outcomes. This valuable learning will further their enlightenment and upgrade their consciousness.

They are expert astral travelers, and some have been known to take astral vacations.

Dimensional Capacity: Pleiadians have multirealm commandership. They are able to reassemble their physical bodies from one location to another through a molecular stimulation of light codes. They can sweep an entire civilization from the ground up into a mothership in less than a second.

Some Pleiadians come to our space-time from the future.

Renegade Pleiadians

Universal Origin: Renegade Pleiadians come from the Pleiades star cluster.

Physical Characteristics: Renegade Pleiadians are powerful beasts, evil Titans standing eight to thirteen feet tall. They manipulated their gene pool to produce warriors built like tanks, and to alter their original hair and eye color to black or dark brown.

Belief System: Renegade Pleiadians believe in service to self. They believe in their dark undermonic laws to gain power, technology, and material wealth. They defile all that is good and righteous, and mock the Cosmic Law of One. They have no clue how their afterlife will be confined to the dark astral universe, imprisoned in oppressive consciousness. If they choose to correct their karma, they will reenter reincarnation cycles. Those new challenging lifetimes will be designed by their celestial sentries.

Cosmic Agenda: Renegade Pleiadians are aligned with and work alongside the Orion Empire. Even though they want to please their reptilian comrades, the renegade giants will someday try to overthrow them, and wipe out as much of their race as they can. Their ultimate goal is to control the universe and rule it with an iron fist.

Angelic Corps psychic detectives monitor the thoughts of highly dangerous malevolent beings in the universe through a cumulative monitoring known as reflectivating. This monitoring is not an invasion of private thoughts; it's heightened intuition. They have monitored the Orion Empire as a whole, and their intuition is that the renegade Pleiadians, who are building up their army in great numbers through genetic engineering births of males, will someday make their move to defeat the Draconians. The Greys are next on their extinction agenda.

Renegade Pleiadian Male

Technology: The renegade Pleiadian fleet is oval-shaped and black. Motherships look like thin wafers with dome protrusions on the underbelly. These podlike facilities are designed as genetic labs, prison cells for captured species, sleeping quarters, food preparation areas, farms, etc. The main floor is the command and weapons center. Inward grooves on the roof of the mothership store the fleet, which are able to take off at a second's notice. These fleet craft can be manned or unmanned.

Consciousness Abilities: The renegade Pleiadian's dark telepathic chatter is aligned to undermonics, the opposite of harmonics. They use a hive mind to interact as a race, to see who can come up with ways to undermine the Galacterian Alignment and its subdivision the Star Seed Alignment, and propose diabolical scenarios to gain control over worlds and outdo the Draconians, while pretending to be their allies.

Dimensional Capacity: The renegade Pleiadians use the dark matter universe for travel, and create temporary dimensional pockets for their craft to take refuge.

Procyonan Male

Procyonans

Universal Origin: Procyonans are from Procyon (Alpha Canis Minoris), the seventh brightest star in the sky, and one star in the Canis Minor constellation.

Physical Characteristics: Procyonans are basically indistinguishable from Earth humans, although their facial features are more pronounced, chiseled—a symmetrical perfection that foretells of grace, strength, and endurance. And men and women are taller than humans overall, ranging from six to nine feet. When in the presence of this race, one feels safe. There is an intensity about them, but in a progressive way. This tenacious trait assures that they solve every problem without deviating from their chosen track. They see everything through to the end.

Belief System: Procyonans follow the Cosmic Law of One. They are cautious of governments, not the people, on dualistic worlds. They are champions of cosmic underdogs.

Cosmic Agenda: Procyonans chart uncharted universal territories for the Galacterian Alignment and participate in the Melchizedek Star Seed Program. On universal space explorations, they are profound spiritualtists (spiritual scientists) that uncover Creation's meaning behind whatever heavenly body they are studying. Their library is extensive and is shared with the Galacterian Alignment.

Technology: Procyonans have Galacterian motherships, starships, thought-ships, fleet ships, and scout ships. All their craft and command centers use biological supercomputers.

Onboard many of their starships and motherships are crystal amplifiers, which assist star seeds to raise their vibration on schedule with the Divine Plan. They also place these crystals in strategic places around the globe for the same purpose, usually in areas where a concentration of star seeds live. These amplifiers also assist the citizens of Earth to elevate their spiritual nature.

Consciousness Abilities: Procyonans are fully conscious and sovereign beings. They are advanced telepaths and can send messages farther than most of the Galactic Kingdom. About half their population has the ability of telekinesis, to move objects with the power of their mind.

During meditation, they focus on the frequencies of sixty-four interlocking tetrahedron crystals of various colors and energies, to take their mind to the invisible worlds. There, they attend school on whatever areas they wish to improve.

All of their technology onboard craft and healing tools are based on a higher vibration and concentration of crystal frequencies that they brought back from the invisible realm.

Dimensional Capacity: Procyonans have multirealm commandership. They like to recharge by visiting the older astral dimensions through mind jumps into that consciousness and bathing in the light there. It's freeing to the mind, energizes the body, and refreshes the soul—a meditation practice that only advanced races are able to execute.

Proxima Centaurians

Universal Origin: Proxima Centaurians are from the binary star system Proxima Centauri in the constellation Centaurus.

Physical Characteristics: Proxima Centaurians are indistinguishable from Earth humans. Females are five foot five to five nine, and males are five foot eight to six five. When gazing into their magnificent eyes, you'll experience their deep understanding of the cosmos. Their bodies are exquisite specimens of the human form. They feel exceptional beauty from within, a feature that makes their whole persona sparkle.

Belief System: Proxima Centaurians believe in the Cosmic Law of One. They are aware that everything vibrates at a different frequency. They are able to tune the frequency of a younger soul to their frequency to temporarily give them a preview of what is to come when their spiritual sustenance increases. This allows the student to work toward their goals on a well-paced and balanced path.

Cosmic Agenda: A member of the Galacterian Alignment, and participants in the Melchizedek Star Seed Program, the Proxima Centaurian race is here to assist Earth in its dualistic proclivity: to spiritually raise, to uplift mentally, to reveal health—all roads that lead to true happiness. On a cosmic level, through the galaxies, they prepare planets in star systems ready for human habitation. They are also master botanists and introduce fauna and flora, especially healing herbs to worlds in need.

They have galactic command centers stationed above and below the Earth.

Technology: Proxima Centaurians have Galacterian motherships, starships, thought-ships, fleet ships, and scout ships. All their craft

Proxima Centaurian Male

and command centers use biological supercomputers. Through imaging technology, they can create physical objects from nothingness, materialize and dematerialize, reverse the aging process, and alter space-time and reality.

Because of their advanced spiritual vibration, they work at the highest tiers of scientific and technological knowledge. Being inside their motherships or craft is like catapulting yourself into a world of never-before-seen multidimensional technology.

Consciousness Abilities: Fully conscious and sovereign beings. The Proxima Centaurian race has extended their fully conscious brain capacity through cerebral cortex amplification. They speak in their own language amongst themselves telepathically, or can easily learn languages in a few hours' time through mind absorption technology. Also, conversationalists can wear an ear translator, which allows them to understand different languages telepathically.

They are able to heal the sick by condensing the smallest particle between the void and the material, and then regenerate the cells to perfect harmony and health.

They have the innate ability to control nature.

Dimensional Capacity: Proxima Centaurians have multirealm commandership. They attune themselves harmonically to the very fabric of the universe, the dimension in which they wish to travel, and match the frequency to their craft. They slide from one dimension into another with ease. Some come to our space-time from the future.

Sagittarian Male

Sagittarians

Universal Origin: Sagittarians are from the Sagittarius constellation, in the southern sky.

Physical Characteristics: Sagittarians look like Earth humans, but women range from eight to ten feet and men from eight to twelve feet. Sagittarians are very angelic, pure, and refined. Like other star nations their skin colors range is extensive. Their eye shapes can be linear or slant slightly.

Belief System: Sagittarians believe in the Cosmic Law of One. They are connected to the unseen universe from which all things flow, and they believe in unity intelligence.

They are geniuses of physical, biological, and psychological sciences, and have the largest widespread departments covering each area. Being scientists, they maintain a healthy spiritual equilibrium, aware that other driven science-based races, like the Draconians and the Greys, lost their connection to Source, and were unable to ascend to the next dimension of consciousness.

Cosmic Agenda: Although their participation is minimal, Sagittarians play a part in the Melchizedek Star Seed Program that is currently focused on Earth. They work alongside other star nations in galactic command centers stationed above and below the Earth.

Technology: As a member of the Galacterian Alignment, Sagittarians have Galacterian motherships, starships, thought-ships, fleet ships, and scout ships. All their craft and command centers use biological supercomputers. All space vehicles travel through time-space funnels and the innerspace continuum. They have one starship that is designed like a giant crystal sphere with long crystalline points. It's quite beautiful to look at, and the interior of the ship is like being inside a crystal city. It's a magical creation. The starship draws its energy from the sun or from starlight.

Consciousness Abilities: Sagittarians are fully conscious and sovereign beings. Although they are expert telepaths, they are exceptional cosmic linguists, and therefore effective diplomats. They are well versed in universal languages, and speak in home worlds' tongues while on diplomatic missions, as some races are more comfortable with the spoken word. They also have a unique way of transferring deep love to others when speaking telepathically and sharing images from mind to mind. This special way of communicating mirrors the Angelic Kingdom.

They are highly developed intellectuals who are experts in profound thinking, strategists of overcoming difficult situations, deep writers and philosophers, and of the human condition across the expanse of cosmic society.

Dimensional Capacity: Sagittarians have multirealm commandership. They have petitioned the Angelic Corps to enter dimensions in other universes for exploration in all areas. Permission is pending.

Blue Sirians

Universal Origin: Blue Sirians are from Sirius B, a white dwarf star, in the Canis Major constellation.

Physical Characteristics: Blue Sirians wish to be perceived for their multidimensional energetic frequency, their physicality; they are tall warriors of the light, with bright blue skin, elongated faces, attractive and appreciative features, and large ears tuned to universal frequencies. They are void of any bodily hair.

Belief System: Blue Sirians believe in the Cosmic Law of One. They are at one with the universe, the heart and mind of Source, the one soul of Creation, from which all beings are interconnected.

Cosmic Agenda: Blue Sirians are superintuitive, ultrapsychic beings that have the innate ability to see the unseen lay lines and grids of the cosmos and planetary fields. They assist worlds in the throes of transformation that are transiting toward and to ascension by stimulating lay and grid lines with Living Light codes, to anchor in these new energetic frequencies, to awaken empowerment codes within every being. They balance planet meridians and soul frequencies during this transformation. They are the universal experts on lay lines and holographic grid templates. They are also a member of the Galacterian Alignment.

Technology: Blue Sirians are the guardians of the ancient spiritual technologies.

Consciousness Abilities: Blue Sirians are fully conscious and sovereign beings. Besides practicing telepathy and coagular wave bending, they are also mental kineticists, able to harness, move, and flow energy from Source through lay and grid lines.

Blue Sirian Being

Dimensional Capacity: Blue Sirians have multirealm commandership.

Human Sirians

Universal Origin: Human Sirians are from Sirius B, a white dwarf star, in the Canis Major constellation.

Physical Characteristics: Human Sirians are indistinguishable from Earth humans. Men range from six foot five to eight feet. Women are six to seven feet. Human Sirian bodies are very well proportioned, with attractive facial features. They are the perfect marriage of physicality and the soul. Their auras vibrate at such a high rate that one can sense their tremendous clarity. They have attained the ultimate human potential in their personal and social life. Human Sirians live in joy, learn in light, and open their hearts to everyone. They empower one another; their race has tremendous self-worth. Their coloring is spectral, sometimes patterned, but most are light skinned or brown. Their eye colors are brown, green, blue, violet, and occasionally a mixture of one or more colors. Looking into their expressive eyes is like traveling through time, gaining access to windows of love in every dimension.

Belief System: Human Sirians follow the Cosmic Law of One. They are conscious of their higher self, of transcendental reality, of the Oneness of All Creation. They are aware that our true state is love, peace, light, and harmony. Some star nations call them blissful spiritual warriors. They align to the universal spirit for spiritual sustenance, mind expansion, energy, and inspiration.

Cosmic Agenda: Some of the greatest Galacterian Alignment science commanders are Human Sirians. They also participate in the Melchizedek Star Seed Program. Their star seeds on Earth are masters at psychic protection, aura amplification, DNA rejuvenation, and energy balancing.

Sirian Male

Imprinted in our mind is the awareness of our soul, our spirit connected to the One Spirit, and once we awaken to the greater reality the Human Sirians will tutor us in the ways of the universe from behind the veil. Once we become fully conscious, they will tutor us in person.

Technology: Human Sirians have Galacterian motherships, starships, thought-ships, fleet ships, and scout ships. All their craft and command centers use biological supercomputers.

They have the same Galacterian healing technologies as other star nations, but they also have created some unique technology of their own. Perhaps the most popular piece of tech is the cellular generator. When they stand on a circular disc that flutters with molecule-like stars, a blue flame rises to engulf the body. As this fire enriches every cell, rejuvenation occurs, and the cells are amplified to a divine resonation.

Consciousness Abilities: Human Sirians are fully conscious, sovereign, and telepathic beings. They have devices that increase their telepathic reception over longer ranges.

They extensively astral travel the astral-verse and sift through layers of consciousness. The megadata is then brought back home to enhance all of cosmic civilization.

Dimensional Capacity: Human Sirians have multirealm commandership. They are able to reassemble their physical bodies or any matter from one location to another through a molecular stimulation of light codes. They can sweep an entire civilization from one location to another.

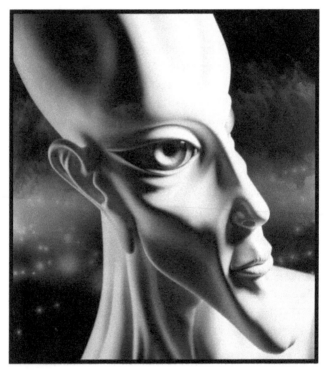

Titan Sirian Being

Titan Sirians

Universal Origin: Titan Sirians are from Sirius B, a white dwarf star, in the Canis Major constellation.

This Sirian race has no name and many names, simply because they now exist beyond identity in form or place. Their image was revealed for our curiosity, an energetic blueprint of their multidimensional consciousness.

Physical Characteristics: Titan Sirian men and women are eight to twelve feet tall. Their elongated heads taper down to an extended thin chin. They have exquisite large almond-shaped eyes that look like glistening aqua blue water jewels. Gazing into them is like being pulled into the warmth of Source, into the womb of Creation, into the love that created all that is. They have long, thin appendages with boney joints, and they move melodically, gently, and elegantly.

Belief System: Titan Sirians are of the One Consciousness, of the universal community, representing the unity of all species and all life, and where polarities finally and ultimately meet within a single circle of Source.

Cosmic Agenda: Titan Sirians are higher realm mentors to the Galacterian Alignment and participants in Melchizedek's Star Seed Program, where they are master teachers to worlds such as Earth.

Technology: Titan Sirians' craft and computers are not physically built; they mentally integrate their minds into the quantum field for travel and information.

Consciousness Abilities: Titan Sirians are fully conscious, sovereign, and telepathic beings. They speak the light essence language. They are closer to light forms and can shift their appearance with a thought. Titan Sirians are not attached to their identities the way humans are. Their minds radiate pure love.

Dimensional Capacity: Titan Sirians are multidimensional beings. They interact between the veils, planting stars of consciousness, concepts of cosmic wisdom and knowledge into the mind gardens of evolving races. They tend to them like their own children, raising them into adulthood, and into their rightful place as sovereign beings.

Was Pharaoh Akhenaten Sirian?

Extraterrestrial theorists suggest that Pharaoh Akhenaten of the eighteenth dynasty is a hybrid with Sirius B DNA. Sirian genetics could have been injected into his embryo for an upgrade in consciousness. That might explain why Akhenaten looked like an extraterrestrial. He shook the foundation of Egyptian religion by erasing all the gods of worship, and turned their thinking to a singular god.

Solar Light Beings

Universal Origin: Solar light beings have attained their status either through multiple ascensions or created status. They have ascended multiple times and are a part of every race; therefore they exist in every constellation. The oldest light beings exist in Paradise, the heart of the central universe also known as Source. It's the geographic center of infinity. In Paradise there is no duality—only supreme love, happiness, and the bond of brotherhood and fellowship.

Physical Characteristics: Solar light beings are pure light and formless, maintaining pure bodies of energy, but if they peeled their garments of illumination away, they would choose the most beautiful and handsome embodiments personifying perfection. Light beings in the forms of other universal creatures would be at the pinnacle of their most pleasing phenotype. Their height can range from six feet to one hundred feet and more. When on a universal mission, they are able to attain any height depending on the expulsion of light from their forms required to seed a galaxy or world with a Living Light program. They are divine creatures.

Belief System: Solar light beings teach that throughout every being's journey their soul gathers light, and being part of the one soul of creation, this light increases the light within Source. It is their great privilege to facilitate learning programs so that every soul may fulfill their life contract on what it is they designed to come into physicality to learn. Some light beings are assigned to mighty messenger star seeds to assist them in preparing and executing universal teachings to the masses.

Solar light beings remind us of the simplest, but most profound truths: Give love, receive love. Treat every day, every moment, as if it were sacred. You've been given the greatest gift of all—life. Enjoy it with your family, friends, and fellow beings. Remember that smiles transform the day.

Male Solar Light Being

Cosmic Agenda: Solar light beings that have ascended far into Paradise, into the Supreme Source of Light, create and write Living Light programs that are dispersed to all worlds in the younger dimensions that outline ascension guidelines. On a more finite level, every being is a spark of the Divine Creator, and more individualized programs are downloaded into every soul.

Technology: Solar light beings, because they are pure light, harness the power of suns and also stars. They travel the universe by jumping from sun portal to sun portal. When traveling through our sun, they can also astral project their light form to anyone on the planet day or night, veiled or unveiled. Some have been sent to reignite the light in a star seed or mighty messenger who has become spiritually stagnant.

When a mighty messenger has spiritually elevated to a new potential, the solar light being will indoctrinate and infuse the messenger with divine fire. (Archangels and angels perform this rite too.) This ritual will always take place in the wee hours. The process is uniform for every mighty messenger; they will wake in the middle of the night lying on their backs, and suddenly a one-foot golden-blue flame sprouts up and out of their heart. This rite allows them access to Source, learning directly from Creation's Divine Thought Forms. The mighty messenger's cognition of universal teachings increases, and they are able to share this knowledge with the masses in layman's terms.

All mighty messengers' lives are engineered by archangels, angels, light beings, Galacterian Alignment officials, and so on. All work is performed to forward the Divine Plan. Some archangels and angels are light beings, but there are also light beings that stand alone in their universal categorization.

Consciousness Abilities: Solar light beings have solar consciousness. They transmute thought beyond the spoken word and telepathy. They encode consciousness through bands of light and deliver them through sunlight to a global consciousness or to an individual. They share with us the nature of consciousness at every dimensional level, and tell us what we have to do to ascend to the next reality and what we have to look forward to in our upcoming elevations. In third dimensional reality they assist us in transforming from dualistic beings to sovereign citizens who are to be fully conscious beings. They hope that when you feel the warmth of the sun, you are thinking of them as they deliver the tools of enlightenment through encoded arms of light.

Dimensional Capacity: Solar light beings exist in the eleventh and twelfth dimensions. Although all ascended solar light beings have experienced every dimension, they are the caretakers of each one.

Soulzars

Universal Origin: Soulzars, like Sirians, are from Sirius B, a white dwarf star, in the Canis Major constellation.

Physical Characteristics: Soulzars are giants, standing twelve to fifteen feet tall. Males have very thin waists with muscular torsos, arms, and legs. Women's waists are thin as well, with voluptuous breasts and slender arms. They are a hairless species and like to decorate their heads with jewelry and/or a starry glowing gel. They move elegantly, almost as if gliding on air. Males are born with blue eyes and females with violet. They also have a hermaphrodite classification, and most interesting is one of their eyes will be blue and the other violet, a balance between masculine and feminine.

They have been bred with human genomes, and their intellect is superior to most races, their large brains like computers retaining vast amounts of knowledge. The Soulzars have been created this way using the best thirty-three human DNA chains from across the universe. Over the next millennia, they will begin a DNA upgrade and adjoin eleven more highly intelligent human races' genomes to their DNA pool, bringing their DNA chain to a total of forty-four races.

Belief System: Soulzars believe in the Cosmic Law of One. They believe in the All, the Oneness of every race, and that our home, the universe, is the most beautiful of homes. They believe in the invisible architect of the universe for this reason: if there is a God, then who created God?

Cosmic Agenda: As a member of the Galacterian Alignment and participant in Melchizedek's Star Seed Program, Soulzars are helping to raise younger worlds into the grandeur of universal society. They like to think of themselves as mothers and fathers to other races who

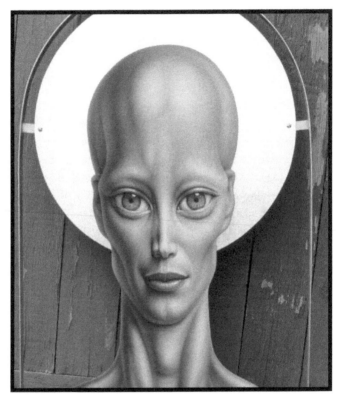

Soulzar Being

can come to them for advice, emotional support, and, most of all, love.

Technology: Soulzars have taken Galacterian technology and modified it. Their mothership designs are two domes interconnecting. The skin of the craft is patina green with golden rivet designs; this is the outer storage facility, or megabiological computer that is also a part of the hyperdrive. Their fleet craft are shaped like hexagons, tetrahedrons, and star pentagons.

The Soulzar's minds are fused with their craft's organic brain. The ship becomes an extension of their body, and their heart and mind the engine.

Soulzars are the healers of healers. They have studied the human condition throughout the universe extensively. Other star nations seek their council for an array of mental or physical conditions.

They are proficient geneticists.

Consciousness Abilities: Soulzars are fully conscious and sovereign beings. They are categorized as cosmic telepaths. Their elevated mental capacity encodes a band of light with a verbal and/or visual message and sends it to the recipient using a interdimensional light delivery system of their own creation.

Dimensional Capacity: Soulzars exist in twelve dimensions, and each density works in unison with the others. The higher dimensional light beings of the Soulzar consciousness can enter lower dimensions in bodily form by entering a soul vector suit, a temporary body designed to inhabit their soul.

Created superangel Gavalia and ascendant superangel Galantia

Superangels

Universal Origin: Superangels are from the Isle of Paradise.

Physical Characteristics: Superangels are androgynous beings.

Belief System: Some created angels in the past believed that ascended beings who achieved status equal to angels did not deserve such universal classification. Superangels work in twos, one ascended angel and one created angel, to set a constant example that both fractions can work together harmoniously.

Cosmic Agenda: Superangels work in union with the Galacterian Alignment. Superb seraphim (superangels) are universal guardians, protectors of higher realms, mentors to administering angels and other angelic sects, and counselors to the redemption angels from the Lucifer Rebellion. It was said of Lucifer, "You were perfect in all your ways from the day you were created till unrighteousness was found in you." Lucifer was a created son and reigned as Satania's System Sovereign (Earth belongs to this system), for more than five hundred thousand years, and then he rebelled against the Universal Father and his vicegerent son, Michael. Lucifer contested Michael's right to universal sovereignty. Lucifer's trial was lengthy. He ultimately lost and was imprisoned for two hundred thousand years on a prison planet. Then, after much reflection, Lucifer was put on trial for soul death. He turned one-third of the Angelic Kingdom against Michael.

During the Lucifer Rebellion, created superangel Gavalia and ascendant superangel Galantia and their team measured the various planetary consciousnesses to calculate a precise number of those who wished to join the rebellion. They, along with the Emergency Melchizedeks, also collected antidotes from other worlds in prior turmoil on how to manage the rebellion.

Superangels also monitor energy fluctuations across the universe and its bandwidth growth. From broadcast stations across the cosmos, they transmit and receive celestial broadcasts, universal information vital to all unified worlds. Citizens on their home worlds gather in outdoor coliseums to hear the broadcasts in a social setting. Interaction and stimulating conversations ensue about a wide range of topics on any given day.

The superangels will again broadcast to Earth once the planet has rejoined universal society.

Technology: The dial on the belts of superangels are used for travel into the spirit or physical worlds. By locking the dial, they are able to maintain any form, for as long as they require, in whatever dimension they work in. They travel in singular or group Merkabah vehicles or in Galacterian craft.

Consciousness Abilities: Superangels have avatar and solar consciousness. They speak the light essence language. Their superangelic consciousness works directly with Paradise Son (Archangel) Michael and other Paradise Sons. Their minds are divine.

Dimensional Capacity: Superangels are masters of the multiverse. With hierarchal permission, they can travel to any of the seven superuniverses or into one of the 700,000 local universes each superuniverse houses.

Synthetics

Universal Origin: Synthetics are biological synthetic creatures; AI with an evil twist. If coming in contact with a synthetic, you would sense it is inhuman—but that is unlikely, as they rarely mingle with humanity.

Physical Characteristics: Synthetic creatures are designed to look human or alien. They're grown in artificial embryo sacs and void of a soul. Souls are born of the Creation, and without the building blocks of divine intent, a soul may not be born into a synthetic. The designers of this technology found a way to attach disincarnate dark entities under their control into the spirit matrix, or to have multiple evils hosting them. From the time they are born, synthetics are bred with the bidding of their programmers. Their auras are extremely dark.

Belief System: Synthetics have no belief system per se; they only wish to carry out the orders of their creators.

Cosmic Agenda: The synthetics are an extension of their dark programmers, used to do their dirty work, including keeping them safe from harm. They usually are internal psychic war agents, sitting in a room and directing dark evil thoughts to a person or group. If directed to an area of land they want to control, they are aware good people will sense the dark vibes and stay away. Their programmers know the power of prayer, so they do the opposite to achieve the opposite result. Synthetic beings could be anything the diabolical mind can conceive.

Technology: Synthetics use biological entity tech with sinister intentions.

Consciousness Abilities: Synthetics are constituted consciousness units. They have speech, but not telepathy.

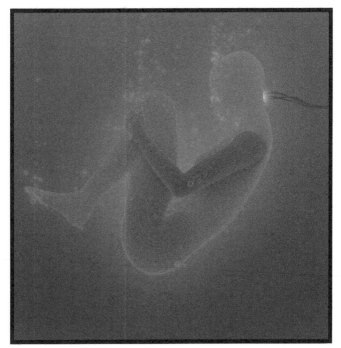

Synthetic Being

Dimensional Capacity: Synthetics are third density, although their programmers can project them into the dark astral realm or into dream states to promote fear in the programmers' enemy.

Ummites

Universal Origin: Ummites are from the planet Ummo, a one continent world with many large lakes. It is near Sagitta, a dim constellation in the northern sky.

Physical Characteristics: Ummites are indistinguishable from Earth humans and Nordic-appearing. Male and female heights range between eight and nine feet. They have a heightened sense of smell and are primarily nocturnal.

Belief System: Ummo's globe is overseen by four people who have been chosen by the populace. There is no monetary system, as with all star nations. Society thrives through a joint effort and discipline. All carry out their duties knowing that if they fail to perform, others in their society will suffer. Their social structure is efficient and unified, using the principles of the One Spirit (the life generator), the glue that holds the physical worlds and its dimensions together.

Cosmic Agenda: In March of 1950, Ummite explorers came to Earth to learn about our cultures. They disseminated scientific information about our universe to selected individuals through letters, including leading scientists, to expand their thinking, to concentrate on planetary problems that required immediate attention. There were also codicils on technical information: technology safe for the environment and educational improvements for all of humanity. The Ummites visit Earth periodically to this day.

They live by the Cosmic Law of One: that every world, its consciousness as a whole, is responsible for its demise or elevation. They shared their wealth of spiritual knowledge with us to help us climb out of the cesspool of wars, pollution, and social unrest. They hope we will heed their warnings and not destroy our race and planet.

Ummite Male

Technology: Ummites have oval saucer-shaped craft. Their ships and world use biological supercomputers.

Consciousness Abilities: Ummites are fully conscious and sovereign beings. They are telepathic empaths. They say that all sentient beings are linked to Creation's soul through man's encephalic cortex.

Dimensional Capacity: The Ummites say the multiverse exists because pockets of reverse time and positioned matter exist in areas where the light velocity is faster. The electrical and magnetic fields and stationary electrical fields do not radiate out of reversed space time. Universes formed out of dark matter exist in reverse time. They invert the dark matter universe(s) to exist in our space time and travel through those corridors across the universe, cutting travel time to a minimum.

Vegan Female

Vegans

Universal Origin: Vegans are from Lyra, a small northern summer constellation.

Physical Characteristics: Vegans are indistinguishable from Earth humans. Women range from five five to six four and men from six to just over seven feet. Skin colors are primarily white, golden, or brown, and as with all star nations, the populace is diverse in skin pigmentations. Vegan women are considered the most beautiful in the universe. They have a unique copper hair color distinct to their culture. Besides the normal range of eye colors, they have a shade of yellow unlike any other star nation.

Belief System: Vegans believe in the Cosmic Law of One. They believe we are all a particle of God, of Source, of the universe. We are all interconnected, and through interaction we the many operate the whole system. We are all points of consciousness that are linked to the universal mind. They also practice the Law of Love. They believe that love flowers from within, and law is how you interact with others. Vegans revere Goddess energy and embrace the divine feminine.

Cosmic Agenda: Vegans are part of the Galacterian Alignment and participate in the Melchizedek Star Seed Program. Their galactic command centers are stationed above and below our planet.

The Vegans have a highly developed psychic attunement with Creation. The goddesses and females of the land focus their healing energies on Earth's heart center. They cater to Earth's soul, knowing she must also heal as the population's consciousness increases to transit to the next phase of evolution. Vegans, as well as all star nations, are aware that planets are living entities. All worlds are revered and tended to with loving care.

Universal citizens travel to Vega to learn the goddess ways and principles, igniting the divine feminine within, and to feel the love of the Cosmic Mother on a grander scale.

Technology: Vegans have Galacterian motherships, starships, thought-ships, fleet ships, and scout ships. All their craft and command centers use biological supercomputers. All craft are able to travel through time-space funnels and the innerspace continuum.

Vegans were influential working alongside other star nations in developing deflector weaponry, a technology that destroys enemy fire in outer space and in ground combat. This technology was developed to save lives on both sides.

Through imaging technology they can create physical objects from nothingness, physically materialize and dematerialize, reverse the aging process, and alter space-time and reality.

Consciousness Abilities: Vegans are fully conscious and sovereign beings. They are expert telepaths and their sensory perception is beyond the cosmic norm. They are great foreseers of the immediate, distant, and far-off futures. They are quite advanced in all areas of life, being direct descendants of the original Lyran colony.

Dimensional Capacity: Vegans have multirealm commandership. They are able to reassemble their physical bodies or any matter from one location to another through a molecular stimulation of light codes. They can sweep an entire civilization from the ground up into a mothership in less than a second.

Venusians

Universal Origin: Venusians are from Venus, located in the Milky Way galaxy. They live in the interior of the planet under synchrotron skies. Venus is the planet that teaches, gives, and restores love unconditionally.

Physical Characteristics: Venusians are indistinguishable from Earth humans. Generally, women are five foot five to six two and men are five nine to six nine. The women are exceptionally beautiful and the men quite handsome, and their eye colors are more vibrant than on other worlds. They radiate love and peace, and they are clear and precise thinkers. Other races, including the Pleiadians, have said, "To be in the presence of a Venusian is to be in the presence of divinity in its true form."

Belief System: Venusians believe in the Cosmic Law of One. They are the master teachers of the very foundations of unconditional love, of the infinite I AM THAT I AM teachings. Their faith is in the ALL; of everything created across the expansive Creation. They believe that everything is unified; that in this sea of energy we are all droplets in the ocean of space, and when our rings of water, our vibrations, reach out to touch and affect those around us, it's either in a positive or negative way, depending on personal intent. They know that the constant evolving mind reconstructs every order of Creation across a network of consciousness that spiritually affects and aligns all forms of life. Therefore, they are responsible for all their actions, and perform their spiritual duties according to cosmic law.

Star nations seek their council in all areas. Venusians are exceptionally loving and caring, and great motivators. But they can be fierce warriors when challenged by antagonistic races. Superb diplomats, they assist star nations to transform conflicts, to recognize

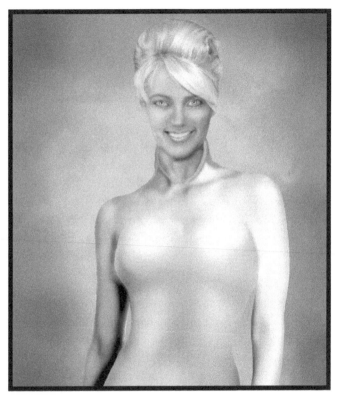

Venusian Female

the light in every situation, and that the only outcome for spiritual elevation is love.

Cosmic Agenda: Venusians are members of the Galacterian Alignment and participate in the Melchizedek Star Seed Program. Since they are our closest planetary neighbor, they will be a major part of First Contact, when Earth again rejoins universal society.

They have galactic command centers located below, on, and above the Earth, in accordance with the Galacterian Alignment, as well as solo bases.

Venusians also help humans understand the meaning of the I AM THAT I AM teachings behind thought. (Learn more about the I AM THAT I AM teachings in the Moramiams entry on page 161.) They have contacted many humans on Earth and taught them about the universe, and granted them a tour of their saucer craft and motherships. They also influence humans indirectly by planting universal morals in their minds to attain a higher understanding of life and all existence.

Howard Menger, an early contactee, was invited onboard a Venusian saucer craft in August of 1956. They gave him a tour of the galaxy, where he witnessed extraterrestrial civilizations on planets and on the moon.

Another 1950s contactee, George Van Tassel, lived in Yucca Valley near Palm Springs, in a secluded part of the desert. He was visited by a Venusian commander named Solganda, who gave George a tour of his craft and then later imparted blueprints to build a domed building called the Integratron. The Integratron's purpose was to facilitate human bodily rejuvenation and time travel.

Technology: Venusians have plasma solar saucers, thought-ships, time-travel ships, spherical-shaped motherships, starship cruisers that are seven miles long and fourteen miles wide, and orb-shaped scout-ships. Onboard craft, their medical healing officers use a body-shaped gloss white clamshell bed that reads the vital statistics of the body's overlays: physical, emotional, mental, astral, celestial, divine. Inside the inner top and bottom of the clamshell there are alternating blue and green plasma-like tubes under clear crystalline glass. These tubes radiate the precise prescription of light to restore a person to optimum health. The beds are usually used for star seeds and others who require medical healing.

Consciousness Abilities: Venusians are fully conscious and sovereign beings. They are expert telepaths, and besides the normal transfer of feelings, they transmit bountiful love to every individual they communicate with. It's a strengthening, a trust that creates a solid bond within their civilization and others.

Dimensional Capacity: Venusians have multirealm commandership. They are able to reassemble their physical bodies or any matter from one location to another through a molecular stimulation of light codes. They can sweep an entire civilization from the ground up into a mothership in less than a second.

Watchers

Universal Origin: Watchers are from Central Control, the nerve center of the universe.

Physical Characteristics: Watchers are indistinguishable from Earth humans, although they are taller. General height for females is five eleven to seven feet and for males is six to eight feet. They are elders of the universe, ascended and created masters, indefinable by age, and have a variety of skin tones. Their hair is usually long and snowflake-colored. Some male Watchers prefer facial hair of various lengths. Other Watchers choose a more youthful appearance.

Belief System: Watchers believe in the Cosmic Law of One. As every being is sovereign, they monitor younger races to protect them from aggressive races that might rape and pillage their lands or cause harm to the populace.

Cosmic Agenda: The Watchers work in unison with the Galacterian Alignment and are sort of the caretakers of worlds. From galactic command centers located within the inner Earth around the globe, they monitor and record the daily comings and goings of Earth's society. They watch over people of interest, including its leaders, aggressive citizens thinking about perpetrating negative acts against humanity, and invited, unannounced, and restricted otherworldly visitors. They send daily reports to the universal hierarchy with recommendations that will not interfere with the evolution of the planetary consciousness and their right to ascend.

If an aggressive race is trying to infiltrate a world secretly or by direct contact, the Watchers report it, and the hierarchy sends a universal emissary/created being to present another choice to the leaders so that the planet does not become enslaved.

Watcher Male

Aggressive races manipulate cosmic law and entrap naïve, vulnerable younger worlds not smart enough to understand these space swindlers are slowing taking over their world. Once a malevolent race has been invited to a world by its leader(s), there is nothing the hierarchal councils can do except wait for an infraction. All the younger dualistic world sees is space technology and other technological advances being dangled in their faces. The end result is a nation's greed at their citizens' expense.

Technology: The Watchers use state-of-the-art biological supercomputers that are living creatures at the avatar level. They are able to record from second to second every living thing on Earth, its soul history and evolution, keeping a watchful eye on the progress of star seeds, planetary births and deaths, and more. In each command center a giant real-time duplicate of the Earth is erected, from which the Watchers monitor atomics, wars, environmental pollutants, nuclear radiation, the depletion of the ozone layer, and toxic waste, among other things. If an area requires immediate attention, reports are made and immediately sent to the hierarchy at Central Control. The cosmic rule of noninterference with younger worlds and how they affect their environment still applies. The exception to this rule includes things that bleed into the universe and adversely affect interdimensional realities and other worlds—pollutants, atomics, etc. Also, the Earth is being groomed for induction into the Galacterian Alignment of Space Peoples and Planets, so there are certain measures and precautions that have to be adhered to.

Consciousness Abilities: Watchers have avatar and solar consciousness. They can send a person-to-person message to any individual to open a nonurgent telepathic message waiting in their consciousness inbox.

Dimensional Capacity: Think of the Watchers as the caretakers of the dimensions and its life.

Part Two

ZETA HUMANS,

GREYS,

AND MORE

EBE Being

EBEs (Extraterrestrial Biological Entities)

Universal Origin: Extraterrestrial biological entities (EBEs) are believed to have been created in the Orion constellation—one of the brightest and best known constellations—and/or Zeta Reticuli II, located in the Reticulum constellation—a small, faint constellation in the southern sky. It is thought that Zeta races' biological materials were being harvested by Draconian reptiles without consent, possibly through abduction and induced amnesia, and used to create EBEs. This is said to have taken place well over a million years ago. A great war broke out between the races. Although they worked out their differences over time, the EBE slave race is widespread across the universe.

Physical Characteristics: EBEs are engineered humanoid automatons, whose thin frames stand three to four and a half feet tall. EBE skin tones are milky white to gray. Their slanted eyes are hauntingly large, and they wear dark lenses over them for protection against the elements. With plain facial features, their noses, ears, and mouths are diminutive. These creatures have no genitals, are androgynous, and look like children. They are chlorophyll-based beings.

Belief System: EBEs have limited consciousness, but are able to function normally and perform tasks programmed by their creators.

Cosmic Agenda: EBEs are usually sent on scientific expeditions in craft, so that their technological superiors remain safe on their home world or in their mothership stationed somewhere in the solar system.

EBEs collect human DNA during abductions on the orders of their creators. They also place "alien implants" into generations of human families for study. Podiatrist Dr. Roger Leir became renowned for his investigations and surgical removal of alien implants from abductees. It was discovered that the implants were cocooned in a biological material so the human body would accept the foreign device. Dr. Leir said that the implants had unique magnetic properties, strange crystalline elements, and emitted radio waves from space. He also noted that the implants had no visible insertion site and that nerve endings surround the implant. He ran lab tests on the alien trackers and reported his findings in two books: *Aliens and the Scalpel* and *Casebook: Alien Implants.*

If captured and interrogated on another world, EBEs may even be linked to their programmers, who speak through the unit, with the capturers none the wiser, so that the alien superiors learn more about the people there. EBEs are an extension of their programmers.

Technology: EBE scout ships launch from their programmer's mothership.

Consciousness Abilities: EBEs have telepathic and telekinetic abilities to unite their minds to the craft when navigating flight.

Dimensional Capacity: Due to the electromagnetic energy enhancements programmed into their auras, EBEs can tune their frequency to become invisible and walk through walls.

The Grey Races

Universal Origin: Greys are a renegade group originally from Zeta Reticuli II (Reticulum constellation) that left their worlds many thousands of years ago. Some of them were captured and genetically altered by groups in the tyrannical Orion Empire.

Physical Characteristics: A Short Grey's average height is four to five feet. Their heads are oversized, and they have thin frames, lean arms and legs, and a gray skin tone. Their large, dark, bulbous eyes are sinister looking and wrap around their heads.

Tall Greys are six to eight and a half feet, have bulbous elongated heads, thin frames, long arms and legs, and a range of skin tones from pale white to gray. Their giant, dark, insect eyes protrude, and beneath the permanent nictitating membrane lenses are huge human-like eyes.

Taller Greys have large elongated heads and small pure black eyes. They are more menacing than their counterparts and stand nine to twelve feet. Their body frames are more human, and the men are muscular naturally. They have darker gray skin tones, and their facial skin is crinkled like crepe paper.

Belief System: The Grey races believe in complete service to self at any cost to other species' free will to satisfy their needs.

Cosmic Agenda: The short Greys lost their spiritual nature over time from inbreeding, cloning, and their pursuit of technology, and they are no longer connected to Spirit. They are desperately trying to regain that union, realizing a bit late in the game that if they don't mend their transgressions, they are doomed for extinction. They are aligned to the malevolent reptilians Orion Group and the Tall and Taller Greys.

Short Greys are responsible for the alien abductions against humanity to upgrade their race. Biologically covered metallic-like implants are inserted into abductees' bodies, those under their bioengineering

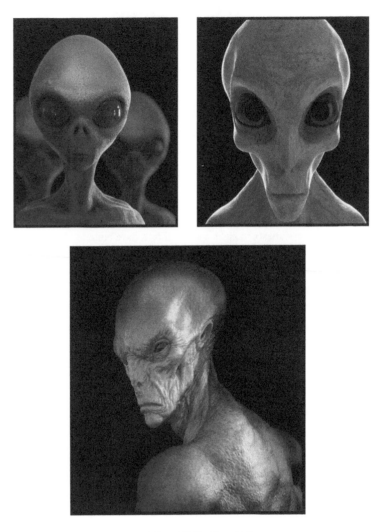

Short, Tall, and Taller Grey Beings

program, to monitor and track them. Through telepathic hypnosis they program their abductees to believe that they are chosen and special.

Against the will of humans, the Greys use our gene pool to save their race that is on the precipice of extinction. They kidnap humans, use mind control to keep them sedated, extract sperm and ova, use female wombs mixed with their genetics as incubators, remove the fetus between two and three months, continue to grow it in their labs, and then repeat the process. They have no remorse for their actions or empathy for their victims.

These three Grey races are diabolical and aggressive creatures. You can keep them away from your vibration by raising your own through spiritual practice, meditation, connection to Source, and shielding your multidimensional bodies with a self-imagined shield of Archangel Michael. They will leave our planet once the spiritual vibration has been raised to an octave that their energy cannot survive in. *They cannot live where there is love.*

Technology: The Grey races' technical sophistication in genetic engineering is superior. Their overall technological advancements are over two thousand years ahead of ours.

The Grey races use hyperspace travel in metallic discuses that are equipped with sophisticated genetic laboratories. Other ships are shaped like a beehive.

Consciousness Abilities: The Grey races have both hive and individual minds. Through powerful mind telepathy, they can control and paralyze humans. They place screen memories in the minds of those they have abducted so they do not remember the horrific ordeals they have had to endure on their medical tables. They are masters of deceit and lies. They possess the power of levitation and can glide across the air.

Dimensional Capacity: Due to the electromagnetic energy enhancements programmed into their auras, the Greys, like EBEs, are able to tune their frequency to become invisible and walk through walls.

Moramiam Beings

Moramiams

Superuniverse Origin: Moramiams come from superuniverse one, known as Flowniningtun. Out of the seven superuniverses, Earth belongs to number seven.

Universal Origin: More specifically, Moramiams are from the planet Morama, Flowloupia universe, from which all creation flows. Morama is the static station of the Infinite I AM THAT I AM. Their world is unique; it cannot be duplicated or destroyed.

Physical Characteristics: The general Moramiam population is eight to nine feet tall, but some can reach heights of ten to eleven feet or more. Moramiams have extra-long elongated heads to house their extraordinary brains. They have expressive, large, almond-shaped eyes, a small human nose, ears, and mouth, and a diamond crest that extends between their brows. In the shallow of the diamond, extra folds of skin vibrate when telepathically communicating. Moramiam initiates wear silver robes, while masters of the realm wear a specific color, or multicolored sun-celled robes.

Belief System: From the Sacred Heart and Mind in the first heaven flows the teachings of the Infinite I AM THAT I AM, the One Soul of God, of Creation, that became the many vehicles of experience. The Moramiams are the keepers of the Flames of Knowledge and Wisdom. This life force of the Creator is transferred to other worlds through encodings in the Light of Fire, the Creator's Language, from the Moramiam priest and priesshood. To learn more about the Moramiams and the I AM THAT I AM teachings, read *The Autobiography of an ExtraTerrestrial Saga* at *www.AutobiographyOfAnET.com.*

Knowledge is based on the three principles of creational thought: thought (the original thought); thought-thought (deliberating on the original positive or negative thought); and using free will to create or terminate the thought (and if created, being responsible for the cause and effect of that thought).

Morama creates, produces, and introduces new life-forms, food, flora and fauna, etc., into young worlds forming, preparing that world for sentient life. The introduction of new foods and plants will nourish the beings and assist in their evolvement on what the creation has set up for them in their new sphere (world) of learning.

Morama's population is comprised of one Eternal of Days representative, Paradise Sons and Daughters, and created beings.

Cosmic Agenda: The Moramiams share knowledge and wisdom and assist the Galacterian Alignment. Some Moramiam souls have stepped down their extraordinary vibratory energy, splitting off fragments of their soul, to incarnate into star nations first, and then into Melchizedek's Star Seed Program to reestablish the I AM teachings on Earth.

Technology: The Moramiams' robes of mastery are celestial computers: the material is ultradimensional, made from components of the universe itself, and houses miniscule suns, a representation of every sun across the vast creation. Each sun collects megadata and transmits the information into the sun cells within their robes, which then is downloaded into Moramiam consciousness.

Consciousness Abilities: Moramiams speak the light essence language. They are the Conscious Ones; they are conscious conduits of the Creation's consciousness.

Dimensional Capacity: Moramiams are cognizant of every dimensional layer across the vast creation. They have the where how to participate in them simultaneously.

Orons

Universal Origin: Orons are thought to be from the Orion constellation.

Physical Characteristics: Orons are between six and seven feet. They are predominately female and female run. Their pigmentation is dark, with rounder heads than the Greys, and with long, extra-jointed limbs, and slender fingers. They may be a hybridization of other insectoid beings, reptiles, and Greys. Their large eyes are bulbous, similar to the Greys. They have nictitating lenses that are used like a pair of sunglasses. Their eyes glow an orange-red color when using night vision or seeing into other realities interdimensionally.

Belief System: Orons believe only in service to self.

Cosmic Agenda: Like many negative races that abduct humans for genetic experiments against their will, Orons make humans feel special to bend them to their will. Human DNA mixed with alien DNA gives Orons an upgrade, as we are considered DNA royalty with ascension capabilities this race lost in their degenerative life force. Another purpose for creating a hybrid race is that when they are a part of us literally, they can study themselves/us more in depth, which helps them to rebuild their new society.

The irony here is that the Orons, through human hybridization, might just become an enlightened race, something they probably did not consider when first mixing their DNA with ours.

Technology: Orons' technical sophistication in genetic engineering is superior. Their overall technological advancements are over two thousand years ahead of ours. They use implants in humans to monitor those under their bioengineering program.

Orons travel in metallic discuses with genetic laboratories onboard.

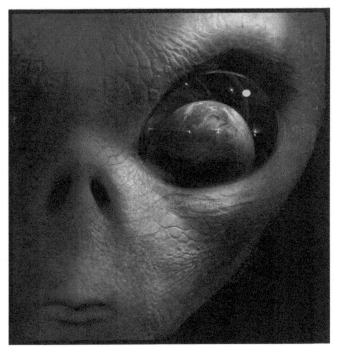

Oron Being

Consciousness Abilities: Orons stare in their abductees' eyes to mind scan them using telepathic hypnosis, gaining access into their subconscious mind to discover their fears and use them against their target. Here they are able to control and paralyze the abductees' bodies while they perform their medical tests and extract genetic materials.

They possess the power of levitation and can glide across the air.

Dimensional Capacity: Due to the electromagnetic energy enhancements programmed into their auras, the Orons (like EBEs and Greys) are able to tune their frequency to become invisible and walk through walls.

Zeta Human Being

Zeta Humans

Universal Origin: Zeta humans are from Cetus, a large constellation in the southern sky.

Physical Characteristics: With thin bodies and bulbous heads, Zeta humans stand about three to three and a half feet tall. Their arms are quite long, reaching below their nonexistent knees, and they have three lengthy fingers and a thumb. They have openings for nostrils, a thin, petite mouth, little teeth, and a hearing cavity on the lower undersides of their heads. Their eyes are twice the size of ours, the irises colored blue, green, brown, or violet. They have a pair of nictitating lenses that cover the whole eye, which acts as a light filter. Think of it as a pair of built-in sunglasses that protect their hypersensitive sight. They also have excellent night vision.

Belief System: Zeta humans believe in the Cosmic Law of One. They enjoy the arts, laughter, and joy. They remind us to make happiness a priority in our lives.

Cosmic Agenda: Zeta humans are part of the Galacterian Alignment, and participate in the Melchizedek Star Seed Program. They, along with Galacterian Alignment star nations, wish to participate in First Contact on Earth. They enjoy studying, nurturing, and developing cultures.

Technology: Zeta humans have Galacterian motherships, starships, thought-ships, fleet ships, and scout ships. Their craft are tailored to their size. All their craft and command centers use biological supercomputers run by their own mind-tech, a variation of the Alignment's mental control systems. They actually make a biological copy of their own brain in personal craft and join their mind with their mind copy for a full sensory experience of flight across the universe. They become the craft and the craft becomes them.

Consciousness Abilities: Zeta humans are fully conscious and sovereign beings. They exceed normal telepathic distances, and love merging their energy with others for a heightened experience of what is being relayed.

Their minds are like supercomputers and hold vast amounts of cosmic information. They not only store every event in their life but also their entire civilization's history, and they can call forth the information at will. They are an extremely advanced mental race—some say it's an overcompensation for their small size.

They also have the ability of telekinesis. It is effortless for them to move any object, no matter how large or small, of any weight, with mental focusing.

Dimensional Capacity: Zeta humans have multirealm commandership.

Zeta Humans– Blue and Golden Caste

Universal Origin: The blue and golden caste of Zeta humans comes from Zeta Reticuli in the Reticulum constellation.

Physical Characteristics: These androgynous Zetas stand four to five feet tall. They have bulbous yet angular heads, slanted eyes with reptilian pupils, and a small nose and mouth. Some of their skin tones are various shades of blue, while others have golden hues—and some are a combination of the two. There are rare cases of blue-skinned Zetas with flecks of gold; these are utterly beautiful, like looking into the depths of the cosmos. They are an introspective race, usually shy, and keep to themselves.

Belief System: The blue and golden caste believe in the Cosmic Law of One. They believe in harmony with nature. They are unconditional love practitioners.

Cosmic Agenda: The blue and golden caste are part of the Galacterian Alignment. They assist scientists on younger worlds via telepathy or through dream states on how to solve problems. They are also proficient cosmic botanists and have quite the library on planetary vegetation.

Technology: The blue and golden caste have Galacterian motherships, starships, thought-ships, fleet ships, and scout ships. All their craft and command centers use biological supercomputers. All craft are able to travel through time-space funnels and the innerspace continuum.

They use healing modalities through small blue-star spheres. By casting one of these orbs into a person's aura, it ignites to cocoon around the body, releasing blue flames that cleanse the mental, physical, or spiritual bodies of any ills.

Zeta Humans of the Golden and Blue Caste

Consciousness Abilities: The blue and golden caste are fully conscious and sovereign beings. Their advanced telepathic nature allows them to stream consciousness in ways that are unique to other races. They secretly code within their messages clues to increasing spiritual sustenance, and telepathic timetables that allow one to connect to others in elevated dimensions for learning.

Dimensional Capacity: The blue and golden caste have multirealm commandership. They are able to travel through the dimensions mentally and absorb learning, increasing their spiritual sustenance.

Zeta Reticulan Being

Zeta Reticulans

Universal Origin: Zeta Reticulans come from Zeta Reticuli in the Reticulum constellation.

Physical Characteristics: A variation of the many species in Reticulum, Zeta Reticulans have slight bodies about four feet tall. They have round heads, and three long fingers with an extra joint and thumbs on each hand. Their eyes are quite large and oval shaped, with huge round irises and pupils colored blue, green, brown, steel blue, or violet. They have nictitating lenses to protect their sensitive eyes. When they converse telepathically, their eyes slant upward, each movement an emotional expression of what they are conveying.

Although they appear of indeterminate sex, they are able to procreate as males or females categorized as hermaphrodites.

Belief System: Zeta Reticulans believe in the Cosmic Law of One. They believe all beings are united with the One Spirit; some souls are awake and serve the universe, while other souls still sleep—slaves to the ego and service to self.

Cosmic Agenda: Zeta Reticulans are members of the Galacterian Alignment. They have the tech to rewire neural patterns to increase brain function in their contactees. After rewiring, their human visitors interface with a screen while wearing neural helmets; this merges their consciousness with the machine. Each invited guest, depending on their abilities, is shown various needs of Earth, and they, if they so choose, are shown how they can be part of positive change. The interface also teaches the visitor their overall program for raising consciousness to repair the damage done to the planet.

There is always a sense of urgency attached to missions that contactees have been assigned, or have taken on of their own accord, to better the world.

The Zeta Reticulans have Earth museums from the beginning of time, showcasing human and extraterrestrial life, and their civilizations. Scenes from various time frames can be viewed like a motion picture. They also store DNA of every past Earth civilization.

The Zeta Reticulans also share that our Mother Planet is a living entity with a soul, and that we are killing her slowly with negativity and environmental pollution. The crucial time is upon us to reverse our ways or Earth is destined to perish—and so will we.

Technology: The Zeta Reticulans have interactive crystal cities that levitate high above the ground, on their home world and onboard motherships. When inside the city, these power stations, encoded with Living Light, assist in extrasensory mind perceptions to access astral realm higher learning. It's similar to returning to Spirit while inside the structures, and drawing on inspiration from the sphere of knowledge and wisdom.

Consciousness Abilities: Zeta Reticulans are fully conscious and sovereign beings. A neuron device strengthens telepathic communication between them and others. Their contactees on Earth are linked to invisible astral headgear that allows the Zeta Reticulans to telepathically communicate with their contactee group as a whole. This is also preparation for when we become a telepathic race.

Zeta Reticulans are masters of mind over matter. They can levitate, glide through the air, and control speed with their thoughts.

Dimensional Capacity: Zeta Reticulans have multirealm commandership. These realm dwellers are so far advanced that they can travel into other dimensions through soul projection or via interdimensional craft.

Zeta Reticuli

Universal Origin: Zeta Reticuli come from Zeta Reticuli, in the Reticulum constellation.

Physical Characteristics: The leader and examiner of the Zeta Reticuli are approximately four and a half to five feet tall. Their heads are larger than their bodies. They have large gecko-like eyes with a vertical slit. Their foreheads are broad, tapering down to a small jawbone. These beings lack body hair, eyebrows, eyelashes, and external ear flaps. Their skin is bumpy and blue-gray to aluminum in color. A membrane covers their toothless mouths, and when they communicate among themselves, it flutters in a singsong *mmmmm* drone. Their narrow palms are cool to the touch, and support three long, slender fingers, one shorter finger, and a thumb.

Smaller Reticulum crew members are approximately three and a half feet. Their head is shaped like a round melon with an enlarged cranial structure. Their large eyes extend to the sides of their face. The iris is either dark, or the pupil extremely large, with barely any white of the eyes showing. Their noses are broad, flat, and petite, with a thin slit mouth. They are sturdy, with broad shoulders and a barrel chest. Their hands have at least four short, stubby fingers and a thumb with no fingernails.

Belief System: We don't know anything about the Zeta Reticuli's belief system.

Cosmic Agenda: Zeta Reticuli are curious to discover how we are like them or different than them physically. They seem to do no harm to those they abduct, wanting only to examine their eyes, nose, throat, bone structure, feet, and hands. They check the nervous system with points on wires. In the famous abduction case of Betty and Barney Hill in the 1960s, the beings inserted a needle into Betty's navel to

Junior, a Reticulum Male

check for pregnancy before we had that technology. If an abductee is fearful, Zeta Reticuli are able to paralyze and subdue them, to avoid harm to either party. If an abductee is feeling pain during an examination, a Zeta touches their forehead for relief.

Technology: Zeta Reticuli have discus craft sixty to eighty feet in diameter, flattened on the bottom and elevated on top, showcasing a curved rectangular hat-shape. It has two rows of five windows that radiate an intense blue-white light. Craft have exam rooms. When in flight it glows a fiery orange-red.

Consciousness Abilities: Some Zetas are more telepathic than others. It's unclear if Zeta Reticuli induce telepathy naturally or through a neural translator. They are able to put humans to sleep by holding up their hand to mentally induce slumber. They are able to erase memories of abduction.

Dimensional Capacity: Zeta Reticuli are most likely multi-dimensional.

Part Three

ANIMAL

AND INSECT

RACES

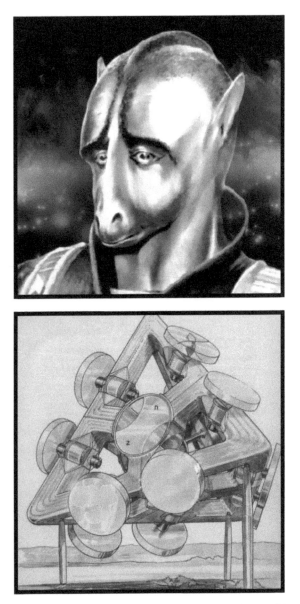

Top: Iargan Male; Bottom: Twelve Universe Synchronization Machine

Iargan Peoples

Universal Origin: Iargans are from the Cygnus Alpha constellation. The planet Iarga is comprised of islands and water. Once water dwellers, Iargans evolved to live on land. The water and land topography of their world are magnificent shades of green. The skies above are cherry blossom pink.

The Iargans' homes are located within oversized cylindrical buildings, each housing ten thousand Iargans, as the islands on the planet add up to the size of Australia, so living areas are compacted. The outer parts of the cylinder are the living units, and the inner courts are manicured parks and gardens. The ceiling is made of glass.

Physical Characteristics: Iargans have small almond-shaped eyes with oval irises and pupils. Their heads are average size, with a boney plate running down the center. They are built for strength, with a broad chest, muscular arms and shoulders, and stocky legs. They are short and compact and walk stiffly. With gray-brown skin, similar to a seal, male heads are covered with short fur-like hair, ranging in color from rust-brown to gold to silver-gray. Women wear their hair at various lengths. Their upper and lower teeth are one solid band of white. Originally amphibians, their hands and feet are webbed.

Belief System: Iargans believe in the Cosmic Law of One and that there is no separation between science and religion. They believe that Earth people and other dualistic cultures throughout the universe must evolve their soul to earn immortality. They further believe that the goal of Creation is to integrate all cosmic communities, every intelligent race, to blend into one unity, one realm of love, one consciousness. Their society strives toward this goal.

Cosmic Agenda: Iargans have come to Earth to share what they have learned about the creation process and various evolutional cycles, and to say that our first cycle is carried out in ignorance, in duality, and the second cycle is in full consciousness. They wish for Earth's isolation from the rest of the universe to be lifted, hence presenting themselves to our world.

Technology: With advanced technological minds, the Iargans are masters of number calculations. They are also masters of acoustic technology, working with sound and sound healing.

Their small (ninety feet in diameter) disc-shaped anti-gravity craft are not capable of flight outside the normal gravitational field of their planet. Motherships are gigantic discuses. The mass kinetic propulsion system is spun in two opposite directional tubes, as matter is forced out at the speed of light but not ejected back into space; it integrates into an anti-mass field, and falls back as immaterial energy in the cosmic carrier field, traveling through the anti-verse.

The analogous visual-mechanical carrier (pictured above) runs in perceptual motion, using no electricity or fuel. It represents the creation of time and matter and synchronizes twelve universes. The machine is several hundred yards long, and operates in perpetual motion.

Consciousness Abilities: Iargans are fully conscious and sovereign beings, able to perform thought transference from mind to mind. A technical device onboard their craft sends messages to a contactee's mind. This method is used so the information being imparted will not affect them emotionally, and will be understood with great clarity.

Iargas work in tribes and think collectively. Group love and friendship are important to them. They describe the high development of the culture in three categories: freedom, justice, and efficiency.

Their law of justice satisfies every being's needs, and thus tyranny of the material possessions is obliterated. When one has everything at their disposal, they no longer desire to obtain wealth. Wealth is equal in every household.

Dimensional Capacity: Iargans are multidimensional. They are able to project holographic pictorials of their home world, other worlds, and other realms from across the multiverse.

Lion-Felines

Universal Origin: Lion-Felines come from the Sirius star system, in the Canis Major constellation. Other feline civilizations are located in the Antares and Lyra systems.

Physical Characteristics: Lion-Felines are comparable to cats and lions on Earth. They have the blended features of a cat and human, with larger eyes and cat pupils. Some have fur, slight fur, or peach fuzz. Both males and females have tails.

The male and female lion/human mix are almost seven feet tall, have manes of hair, more characteristic of humans in bodily form, with a soft chamoix-like flesh in tones of orange, gray, black, and white. Their facial features are more human, but they have cat eyes, cat noses, and smaller mouths. They also have tails. Both are benevolent, spiritual, and intellectual.

Belief System: Feline-Lions believe in the Cosmic Law of One. They believe that every being across the universe is interconnected and affects the chain. So they live their lives in love, affection, championing one another, and other races.

Cosmic Agenda: The Feline-Lions are part of the Galacterian Alignment and participate in the Melchizedek Star Seed Program. They wish to assist us in learning about all cosmic life that is different than ourselves, to learn the customs and cultures of our most unique brethren. They long for the day when they are able to interact with us. When star seeds first wake, they usually begin seeing images of their home cosmic peoples, and then the Feline-Lion race. They want us to know how much they love us.

Technology: Feline-Lions have Galacterian motherships, starships, thought-ships, fleet ships, and scout ships. All their craft and command

Lion-Feline Being

centers use biological supercomputers. All craft travel through time-space funnels and the innerspace continuum.

Their mental abilities are supernatural; they can see into other realms, differentiate between the various dimensions, and sense when an invisible being is present.

They are natural-born healers through touch. They are able to read auras and repair or enhance them with mental focus.

Consciousness Abilities: Feline-Lions are fully conscious and sovereign beings. They are highly telepathic. They have a paranormal sense of intuition. They can perceive danger well before it happens, receiving images and pictures of what is to come.

Dimensional Capacity: Feline-Lions reside in the fourth to sixth densities, but can lower their vibration to interact with races in the third density.

The Mothman

Mothman

Universal Origin: Mothman's origin is unknown, but speculation is that he is a prehistoric subterranean creature. Many wonder if he's an escaped or abandoned hybrid created by extraterrestrials or government geneticists. Others believe him to be an extraterrestrial because the Men in Black have visited over one hundred witnesses to silence them.

Physical Characteristics: Mothman has a powerfully built humanlike body, about seven feet tall, with gray to black coloring, and pterosaur-like wings with a wingspan up to fifteen feet. He has a small, round head, a small nose and mouth, and glowing red eyes, possibly used for night vision. Some say he can fly up to ninety-nine miles per hour. When walking on two feet he totters strangely.

Belief System: Even though his frightening appearance has traumatized many eyewitnesses, and he's believed to be malevolent, no one knows his true intentions.

Cosmic Agenda: It is said that the Mothman is a dark omen of impending worldly disasters; a sign that something horrible is about to happen. Or is he the watcher of astral territories? Does he have the power to grant or deny access between the realms he occupies?

The Mothman first appeared in Point Pleasant, West Virginia, from November 1966 through December 1967. The first Mothman sighting happened on November 12, 1966. Some men were digging a grave for one of the men's father-in-law, at a cemetery near Clendenin, West Virginia. A giant flying man-like bird flew from the trees and over their heads. Roger and Linda Scarberry and Steve and Mary Mallette, two young couples, spotted a huge gray creature that had red glowing eyes a little over a month later. They said it looked like a flying man with a giant wingspan.

Technology: The mothman's technology capabilities are unknown.

Consciousness Abilities: There have not been reports of telepathic communication with the Mothman. He is said to have a highly intuitive sense.

Dimensional Capacity: The Mothman is thought to be an interdimensional being.

Beesonites

Universal Origin: Beesonites come from the Boötes constellation in the northern sky.

Physical Characteristics: Male and female Beesonites are on average five five to six feet. They have a thin insectoid body, four long, slender fingers with three joints and thumbs, enlarged aqua bee-like eyes, a human nose and mouth, and the ears of a small dog. Their skin is tan and smooth, like soft leather.

Belief System: Beesonites are master teachers of the Cosmic Law of One to other cultures who have just made the transformation from duality to fully conscious beings. Their energetic connection between all living things holds a respect like no other. They are known as benevolent sensitives. They are aware that all beings carry the entire history of the universe in their cells. Our interconnectedness to everything creates the Living Library of Akasha.

Cosmic Agenda: Beesonites are part of the Galacterian Alignment and participate in the Melchizedek Star Seed Program. They are also stationed in Galacterian Alignment motherships above Earth and across the solar system in preparation of its dimensional shift.

Technology: Beesonites have Galacterian motherships, starships, thought-ships, fleet ships, and scout ships. All their craft and command centers use biological supercomputers. Some of their spherical fleet and mothership's skin are designed after their own hypersensitive bee-like eyes. The skin serves as a visual amplifier to pull images from afar closer just by thinking it so.

They have the normal healing modalities of the Galacterian Alignment, and incorporate some of their own unique technologies. With third eye sight they have the uncanny ability to see into body overlays

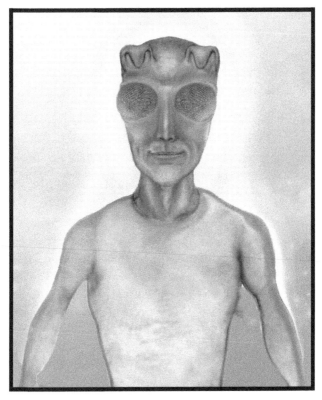

Beesonite Male

and repair the damage through DNA, with the use of the vibrational DNA language converter. The device is a white wand with an oval tip. They understand that DNA is communicative, and when speaking its language, it responds to healing instantly. For more serious cases, they have a DNA rejuvenation chamber that works on the same principles.

Consciousness Abilities: Beesonites are fully conscious and sovereign beings. They use multiwave telepathy, and their mind-to-mind messages are so harmonious that waves of background sounds, music, love, and movement accompany conversation, as if watching a motion picture. It is a most pleasurable sensation when conversing with a Beesonite.

They are able to speak to any energy that is incapable of speech, including inanimate objects. As expert telekinesis technicians, Beesonites are able to fold their mind interdimensionally and program the energy from there to lift any size object in the third dimension.

Dimensional Capacity: Beesonites have multirealm commandership.

Mantis Female

Mantis Beings

Universal Origin: Mantis beings live in various interdimensional folds of space. They exist in the Andromeda and Orion constellations, and others beyond those that humans can't identify.

Physical Characteristics: Mantis beings are seven to ten feet tall, and the ones in the astral world can be as large as a football field. The various varieties are scattered throughout the universe, but all have the basic mantis faces and bodies akin to their counterparts on Earth, except they stand upright. Their faces are long and triangular shaped with large slanted upright black eyes, a sharp pointed beak and two slits for nostrils. Their arms and legs are extra long, with three-pronged hands and feet that have pads on the bottom for grasping. With an extremely bony exoskeleton, their long necks are able to bend at impossible angles. They are an authoritative race, delegating duties to others onboard their craft. They communicate using clicking as well as high-pitched sounds.

Mantis beings have emotions, but because they are unable to use facial expressions like humans do, some might think they are detached and unfeeling. One would have to be telepathic to connect to their interior world to feel and sense their emotions. They are very dedicated and serious when performing their cosmic duties onboard craft, but they are also playful and childlike (but not childish).

Belief System: Mantis beings believe in the Cosmic Law of One. Their lives are dedicated to the Living Light frequencies. Their intention is to hold that fluid frequency in order for all beings to align to universal unification.

Cosmic Agenda: Mantis beings are part of the Galacterian Alignment and participate in the Melchizedek Star Seed Program.

They are more drawn to the feminine, the mother principles of the universe. If this race had a dynasty, there would be a queen; but instead they revere men and women as equals.

As in all races, there are those of a higher frequency and those of a lower frequency. Some Mantis reports state they have been seen in the background of abduction cases, alongside Greys either facilitating or protecting. They harvest human DNA, study our genetics, and examine the human soul in many facets, including reincarnation cycles.

Technology: Mantis beings have Galacterian motherships, starships, thought-ships, fleet ships, and scout ships. All their craft and command centers use biological supercomputers. There are modifications on Galacterian craft to accommodate their limbs, but most of the controls are telepathically operated. They can initiate out-of-body experiences in humans.

Consciousness Abilities: Mantis beings are fully conscious and sovereign. Besides being telepathic, they have a hive mind by choice, and a singular mind. They can enter their species' hive mind and exit at will. While in the hive mind, they can converse with others, share experiences, expand upon teachings, and more. Think of it like a mind-internet. Their minds are extremely powerful.

Mantis beings look at the universe as a symphony, and all the players its instruments. They regard themselves as frequency aligners, assuring all the instruments are tuned. They can hear and sense discord in sectors of the universe, and send their frequencies to those areas for realignment and healing.

They love to astral travel, visiting universal temples of learning in every dimension. When a group of Mantis beings are ready to transit from the sixth dimension to the seventh, the 6D beings astral travel to the 7D learning centers and prepare themselves for the next transition of their conscious development.

Some higher frequency Mantis beings speak the light essence language.

Dimensional Capacity: Although Mantis beings exist from the sixth to twelfth dimensions, they are able to step down their energy to appear in the third, fourth, and fifth dimensions as well. They use crystals, which assist them when transitioning from one dimension to another. Because Mantis beings are one of the most ancient and highly evolved races, they work in realms and dimensions that humans cannot begin to understand.

Part Four

NOTABLE

HYBRID

RACES

Grey Insectoid Female

Grey Insectoids

Universal Origin: Grey Insectoids come from Mu Draconis, a binary star in the Draco constellation.

Physical Characteristics: Grey Insectoids are a hybridization of Tall Greys and Mantis beings. They're eight to ten feet tall, and have long, emaciated-looking bodies, spiny arms, and extra-long, slender fingers. Their oversized bulbous eyes are a cross between the Tall Greys and the Mantis beings' eyes. Some have the mantis pupil, but others do not. Though they look frail, they are quite strong.

Belief System: Grey Insectoids are primarily in service to themselves. They are a neutral yet secretive race, and keep to themselves.

Cosmic Agenda: Grey Insectoids are proficient geneticists. They collect genetic materials from races they come in contact with. Their intention and role in genetic engineering on Earth is unclear.

Technology: Grey Insectoids have advanced cube-shaped craft that are equipped with state-of-the-art genetic labs. They use multi-dimensional instruments when extracting genetic material to avoid pain of those under their abduction. Some craft are organic and others are not. But in both classifications they use mental connections to maneuver their ships.

Consciousness Abilities: The telepathic range of Grey Insectoids exceeds normal ranges. They are a telekinetic race.

Dimensional Capacity: Due to the electromagnetic energy enhancements programmed into their auras, Grey Insectoids can tune their frequency to become invisible and walk through walls.

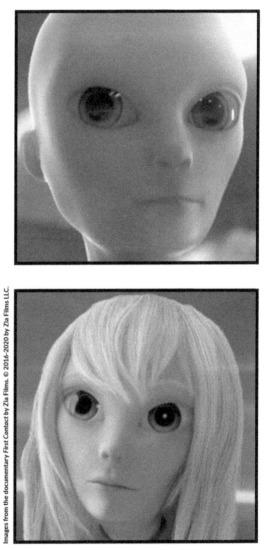

Sassani Male and Female

Sassani Beings

Universal Origin: Sassani beings are from a parallel reality three hundred years in the future. The humans of that version of Earth lost their way in their dimension and destroyed their planet. They mutated into the Greys to survive but could no longer reproduce and were dying out. Since our DNA was close to theirs, they visited our timeline for genetic materials and crossbred with us to create the following hybrids that would allow their culture to continue:

- Hybrid 1: Small Greys are known as the Maz'e.

- Hybrid 2: Taller Greys are known as the Mazani.

- Hybrid 3: Sassani beings are half human, half Grey.

- Hybrid 4: Sha'ya'el beings are slightly more human-looking.

- Hybrid 5: Ya'ya'el beings are very humanlike.

- Hybrid 6: Earth humans becoming the sixth hybrid race known as the E'nani'ka.

- Shalanaya: Hybrid children who will eventually live on Earth with humans.

- Hybrid 7: Anu-Het, a blending of all six hybrid races in a thousand years.

Physical Characteristics: Sassani beings are five feet tall on average. They have thin bodies with larger heads and eyes but smaller ears, noses, and mouths. Their skin is pale. Males have no hair. Females have hair that tends to be white, but there are exceptions. They have five fingers and five toes. They're superintelligent beings with an increased brain capacity, and are emotionally balanced.

Top: Sassani E'ssassani Scout Ship
Bottom: Sassani E'ssassani Landscape

Belief System: The Sassani live by the five laws of creation:

1. You exist.

2. Everything is here and now.

3. The One is The All and The All are The One.

4. What you put out is what you get back.

5. Everything changes except the Laws.

Cosmic Agenda: Because of the successful hybridization of the Greys and humans of Earth, the Sassani feel a kinship to us and have traveled to our parallel reality to share the advanced laws of physics with us, to assist us in moving into the fourth density to the fifth dimension. They have chosen to come and help our transition to be a joyful one, by expanding our awareness of the unlimited possibilities that are available to us.

A simple Sassani belief: *Integrity = behaviors aligned with thoughts aligned with emotions aligned with positive beliefs.* They believe that self-empowerment is our key to mastery.

Technology: Sassani beings have triangular parallel reality travel craft, dark metallic in color, about thirty feet on each side. On the truncated underbelly points there are three blue-white lights, and one orange-red light mid-center. Their smaller craft are scout ships. Slightly larger triangular ships are exploratory craft, and motherships or city-ships are cylindrical and can be one to several miles in length, holding hundreds of thousands of individuals.

The ships are sentient AI that are telepathically connected to the pilots for navigation. They can use electromagnetic and gravitational fields for sub-light travel and use location displacement drives powered by universal energy for instantaneous relocation between star systems.

Consciousness Abilities: Sassani beings are fully conscious, sovereign, and telepathic. They can transcend time and space with their minds and are thus able to contact beings in other parallel realities since they understand that all realities exist simultaneously. This interdimensional communication process is the basis for their ability to channel through Earth humans and other beings since it's nothing more than a telepathic link between two beings in different, but simultaneously existing, dimensions.

Dimensional Capacity: The Sassani live in a different frequency of reality in a higher, fourth density/quasi fifth density reality that vibrates ten times faster than our reality. Thus, while we may think of them as being three hundred years ahead of us in the "future" relative to our reality, their civilization is actually three thousand years ahead of ours technologically and socially.

Bashar of the Sassani Race

Darryl Anka began channeling Bashar, a multidimensional being, and a representative of the Sassani race that lives in a parallel reality we perceive as the "future" and is Darryl's "future" self. Bashar covers a wide range of topics: personal transformation, the history of the universe, extraterrestrials and UFOS, parallel realities, spiritual ascension, healing methods, future technology, sacred geometry, and more.

Seventh Ray Race

Universal Origin: The Seventh Ray race comes from the Pleiades star cluster and the Mira System.

Physical Characteristics: The Seventh Ray race is bred with half Earth genes and half star nation genes. They are indistinguishable from Earth humans, but they are slightly taller. Men range from five nine to almost seven feet, and women five four to six six. Male physiques are slim to muscular, and women are slim to voluptuous. Their eye clarity and color are vibrant. Their auras are so lively that the invisible glow around them can be felt.

Belief System: The Seventh Ray race believes in the Cosmic Law of One. They are spiritually enhanced, bred with divine purpose. They take their roles in the universe seriously. They are intelligent, spiritual, joyful, fun, egoless, and peaceful; all the best human qualities are a part of their genetic makeup. They practice a unified field of trust, truth, and unconditional love.

Cosmic Agenda: A member of the Galacterian Alignment, and part of Melchizedek's Star Seed Program, the Seventh Ray race is a hybrid race of star nations and Earth's gene pool extracted from incarnated star seeds, a part of their cosmic contract. These special children are raised by communal parents.

In young adulthood they are groomed to become fully conscious. Once they reach their teenage years, training begins on activation of their five divine dormant chakra centers above their head, for integration within the seven chakras anchored in the body to create a new chakra system: the Yod Spectrum. This is the juncture when one becomes fully conscious. The individual is now connected to their universal brethren for further edification on the cosmic courses of life. The physical body is now able to reconstruct itself cell by cell with

Seventh Ray Race Male

Living Light to transform itself into a divine body. Once their full Yod Spectrum has been activated (all twelve chakras), and they are fully conscious, they are schooled at the University of Melchizedek in cosmic knowledge and wisdom and of Earth's historical past.

When Earth makes the dimensional consciousness shift, the Seventh Ray race will be assigned to sectors on the planet to become master teachers for pods of people. These cosmic stewards will help our new society adjust from dualistic freedom to the fully conscious unity of the universe. These special beings will be familiar to us, because they are us, and who we will ultimately become.

While they wait for our transformation, they work with star seeds to prepare them for telepathy, the universal way of communicating when fully conscious.

Technology: The Seventh Ray race has Galacterian motherships, starships, thought-ships, fleet ships, and scout ships. All their craft and command centers use biological supercomputers. All craft are able to travel through time-space funnels and the innerspace continuum.

Consciousness Abilities: The Seventh Ray race is fully conscious, sovereign, and telepathic.

Dimensional Capacity: Seventh Ray beings are also master teachers of traversing dimensions, and will share with the students of Earth what they have to look forward to once they reach those densities. They have hypersensitive seer abilities to peer into other dimensions, sift through evolutionary consciousness, and make calculations if heading down a dark, light, or neutral path.

Sirian Insectoid Female

Sirian Insectoids

Universal Origin: Sirian Insectoids are from the Sirius star system in the Canis Major constellation.

Physical Characteristics: Sirian Insectoids range in height from seven foot five to eight two. These insectoids have ultrathin bodies with long arms, four long fingers with an extra joint, and no thumbs. Their lengthy legs are spiny, and under their four long toes are suction-like fibers used for balance. They have quite elongated craniums and a thin facial structure, with large, almond-shaped eyes, usually purple or blue, and a small nose and short, thin lips.

Belief System: Sirian Insectoids believe in service to self and partial service to humanity. They're a neutral race that sometimes disobeys cosmic law. The Star Seed Alignment is working with Sirian Insectoids to bring them into full consciousness and ultimately become indoctrinated into the Galacterian Alignment of Space Peoples and Planets.

Cosmic Agenda: The Sirian Insectoids live on a world of insects that come in many shapes and sizes. They specialize in insect genetics.

Technology: Sirian Insectoids' everyday travel is in oval globes.

Consciousness Abilities: Sirian Insectoids are dualistic beings. Their language sounds like tonal buzzing—and although they are not yet telepathic, they are tonalpathic—able to send, bounce, catapult language tones to a recipient across great distances.

Dimensional Capacity: Sirian Insectoids are moving into fourth density.

Zeta Lizard Humans

Zeta Lizard Humans

Universal Origin: Zeta Lizard Humans come from the Boötes constellation.

Physical Characteristics: Average height for male and female Zeta Lizard Humans is four to five feet. Although they look similar, they are able to tell each other apart due to telepathic honing, identifying each personality internally. They have large pear-shaped heads, enlarged almond-shaped eyes, regular eyelids, and an additional nictitating lens that works like a pair of sunglasses. They have reptile slits for pupils, small open cavities for ears, and a tiny nose and mouth. Their skin is akin to a Gila monster, its hues tan, brown, and beige.

Belief System: Zeta Lizard Humans believe in the Cosmic Law of One. They love their interconnectedness to all beings, to the heart and soul of the universe, to animals and plant life, and so on. They are in tune with nature, and therefore nature is in tune with them.

Cosmic Agenda: Zeta Lizard Humans are members of the Galacterian Alignment. They are onboard Alignment vessels over Earth, assisting with the great awakening on the planet from afar. They enjoy assisting Earth's cultures by elevating our minds, bodies, and souls to ultimately become at one with the universe. They are master diplomats, theologians, and possess immense joy in their heart; three traits that guarantee success while mediating otherworldly disputes.

Technology: Zeta Lizard Humans have Galacterian motherships, starships, thought-ships, fleet ships, and scout ships. They have a variation of fleet craft that are oval, saucer, and cylinder-shaped. All their craft and command centers use biological supercomputers.

Consciousness Abilities: Zeta Lizard Humans are fully conscious and sovereign beings. They are harmonious and empathic telepaths; you feel uplifted when receiving their imagery, emotions, and words and cannot help but smile when they communicate.

Dimensional Capacity: Zeta Lizard Humans can exist up to the fifth density.

Part Five

REPTILIANS

Dinosaurian Male

Dinosaurians

Universal Origin: Dinosaurians are from Bellatrix, the third-brightest star in the Orion constellation. Some Dinosaurians live in the inner Earth, where the origin of their DNA originates. There are many other Dinosaurian races that originated from across the universe.

Physical Characteristics: Males have a height of about six feet, and women are five five to six two. Their dark green skin lightens to beige around the torso. They have bulbous heads; large, round reptile eyes with binocular vision, usually yellow; a short snout; and a wide, lipless mouth. With lengthy spiny arms and three long clawed fingers, they balance on three pronged feet, also clawed. Their movements are jerky and they are fast distance runners.

Belief System: Dinosaurians believe in service to self.

Cosmic Agenda: Dinosaurians serve the Orion Empire. These malevolent creatures' purpose on Earth is to intimidate missionary star seeds, stop the progression of their spiritual elevation, and inject fear into them in hopes they will terminate their assignments. They also infiltrate warped human minds to plant seeds of disinformation—that the nature of the universe is malevolent. These are complete untruths.

Technology: As members of the Orion Empire, Dinosaurians have access to Draconian craft.

Consciousness Abilities: Dinosaurians are able to perform telepathic projection: they are crafty mind technicians, able to enter and poison minds with fear, hate, and intimidation.

Dimensional Capacity: Dinosaurians are third-density beings, but can mind project into other realms.

Draconian Male

Draconians (Dracs)

Universal Origin: Draconians are from the Orion, Sirius, and Draco constellations. The leaders reside in Orion.

Physical Characteristics: Draconians are cold-blooded creatures whose skin is made of scales, scutes, and bony plates. Their most common coloration is brown, beige, and green, or all one color. Others are more vibrant, combining one or more hues: red, orange, yellow, green, aqua, blue, and purple. They are built to fight and have impressive muscular bodies. Their clawed fingers and toes have four digits and a thumb; other subspecies have three fingers and a thumb. Males have distinct mating claws on each thumb used to subdue females and induce pleasure in their mate. Reptile eyes are usually green or blue, but sometimes are orange, yellow, or brown. Males are seven to nine feet and females are six to seven feet. They have a foul, pungent smell, an odor that is designed to weaken its opponent in battle, or to repel enemies. Draconians are not emotional creatures like humans.

Belief System: Draconians are staunch believers that they were residents of the universe first and that it belongs to them. They feel superior over every race, especially human. They teach their children from youth that humans are the dreck of the universe. They are not emotionally attached to their young, and let them fend for themselves. If the child dies due to circumstance, they believe the child was weak and deserved death. Their superiors are the Alpha Albino Royal Dracs. Other reptilian species, their subordinates, know their classification, but to gain power and status they will fight for a more prevalent spot in the clan.

Cosmic Agenda: Draconians want to completely rule the universe. They are the fiercest tyrannical warriors known to any civilization. Fear and weakness are aphrodisiacs to them. They are hell-bent on conquest and control. Excellent navigators of universal territories, they have conquered many planets secretly from the inside. Master manipulators and puppet masters, due to their billions of years of evolution, they alter less evolved civilization gene pools to dumb them down and control them from afar for human food, resources, and underground territory, the population unaware that they are being controlled. Draconians alter the perfection of the human body, including the mind. Hence, thought patterns become degenerative, and the humans' right to ascend into the grandeur of the universe becomes inactive. Draconians do not look for immediate results; they plan everything out carefully to guarantee success. Time is their friend. They are said to rule over Earth from behind the cosmic curtain and are said to have three bases on our planet.

Draconian imperialistic agendas are threatened as we explore the universe, so they try and stunt our growth at every turn.

As with all races, some Draconians are of a lower or higher frequency. All Draconians are not evil, and some have escaped their worlds to live their lives in peace elsewhere, or have ascended. The enlightened Draconians want us to remember to judge the individual, not the race.

Technology: Draconians' ships are disc- or box-shaped. Although their ship technology is organic, it is programmed with dark consciousness and evil intent.

Consciousness Abilities: Draconians shape-shift their consciousness into other dimensions, including the astral plane, and project their consciousness into a body they want to control. They pick lower vibrational humans whose auras are compromised due to smoking, drinking, and abusing drugs. These holes in the auric field give them a point of entrance.

When impregnating human women, Draconians camouflage themselves as handsome human men having sex with them in a dream or altered state. The Dracs then harvest the fetus within two to three months for their breeding program. Female Dracs do the same with human men.

When abducting humans, they implant false screen memories in the host. They can manipulate a mind or body in a dream or in the astral plane, and also attack them while there, instilling tremendous fear upon waking. The Dracs feed off the trauma, a sustenance that gives them power and boosts their ego. They can speak mind to mind, but without the transference of true feelings, as in benevolent races.

The Draconians can control lower vibrational astral bodies from the astral realm. They are puppet masters to those of their choosing within their own race and other species.

Dimensional Capacity: Draconians are able to hide their craft in dimensional time pockets, secret realms not known to man, but known and watched by the Galacterian Alignment. They also have their own travel corridors in the antimatter-universe, black worm-like holes, when trekking between galaxies.

Alpha Albino Royal Drac Male

Alpha Albino Royal Dracs

Universal Origin: Alpha Albino Royal Dracs hail from the Orion constellation.

Physical Characteristics: The Alpha Albino Royal Dracs stand anywhere from ten to twenty something feet. They are monstrous creatures. They are winged, with arms and clawlike hands, and have powerhouse bodies and muscles of steel. Their pure white scaly skin is a combination of scales and scutes. Dracs have small round eyes that are either burnt yellow or orange with reptile slits, and they have bony crests in place of eyebrows. Even though their noses are small, they snort a fierce sound, one of domination. Their lips are thin and wide, covering razor-sharp teeth. They are royalty among their caste, their leaders coming from a long genetic line of rulers.

Belief System: Albino Dracs taught the Draconians, their subordinates, possibly created by the Albinos to do their dirty work, that the reptile community was the first race in the universe and that they own it. Tyrannical rulers, they divide and conquer worlds. They teach their young from the time they can speak that humans are the dreck of the universe.

Cosmic Agenda: Alpha Albino Royal Dracs want to turn as many species as they can to service to self. When they become dark enough, they fold them into their Orion Empire. Alpha Albino Royal Dracs and the Draconians are said to rule over Earth from behind the cosmic curtain.

Technology: Draco ships are blue-black spheres that look like the hump of a black widow. Their motherships are fierce-looking winged creatures. Although their ship technology is organic, it is programmed with dark consciousness and evil intent.

Consciousness Abilities: Alpha Albino Royal Dracs most of the time leave controlling humans to their Draconian brothers and sisters.

Their telepathic abilities among their caste are used along with militaristic hand gestures, and when giving a command to a subordinate the tone of the message and gesture conveys what will happen to them if they do not succeed.

Dimensional Capacity: Alpha Albino Royal Dracs are able to hide their craft in dimensional time pockets, secret realms not known to man, but known and watched by the Galacterian Alignment. They also have their own travel corridors in the antimatter-universe, black worm-like holes, when trekking between galaxies.

Dragon Dracs

Universal Origin: Dragon Dracs are from Alpha Draconis, a star in the constellation of Draco.

Physical Characteristics: Dragon Dracs stand a monstrous fifteen to twenty feet high. Some have been known to grow to thirty-plus feet. Their shiny black reptile skin is a combination of scales, scutes, and horny, boney plates. They are horned and clawed creatures, having three fingers and a thumb and three pronged toes. Their elongated snout, similar to a crocodile, has two rows of sharp gray teeth. Their double wing span is menacing, like evil power being magnetized into dark sails. The Dragon Dracs have slitted pupils and cold, deep blue eyes.

Belief System: Dragon Dracs believe in complete service to self at any expense, and they will kill or ruin anything that gets in their way. They believe they are the dragon gods that rule the universe, so the Galacterian Alignment and the Angelic Corps have a close eye on them. If they are brought up on charges for an infraction against cosmic law, the Alignment and Corps lock the Dragon Dracs' auras to holding cells on prison worlds until they are brought to trial. Many convicted Dragon Dracs (Alpha Albino Royal Dracs and Draconians included) have been put to soul-death for their heinous crimes.

Cosmic Agenda: Dragon Dracs work in secrecy and are in league with the Alpha Albino Royal Dracs, as they seize world after world and fold them into the Orion Empire.

Technology: Dragon Dracs travel in dark ark motherships, tooled to the frequency of undermonics, and are undetectable when sliding into a third-dimensional sky of a younger, less technological world—even when monitored by the Galacterian Alignment. They also have

Dragon Drac Male

their own travel corridors in the antimatter-universe, black worm-like holes, when trekking between galaxies. Although their ship technology is organic, it is programmed with dark consciousness and evil intent.

Consciousness Abilities: Dragon Dracs rule from behind the cosmic veil, and command the Orion Empire in secret through their elusive and mysterious mind powers. Their dark consciousness has a hold over the Draconians and its subordinate hybrids, those polarized to negativity.

Dimensional Capacity: Dragon Dracs are interdimensional sliders. They slip from dark-verse to dark-verse, and create temporary verses, hidden sanctuaries, to escape the Galacterian Alignment or the Angelic Corps. It's a game of cat and mouse to them.

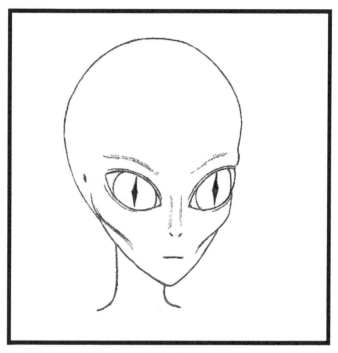

Zeta Drac Male

Zeta Dracs

Universal Origin: Zeta Dracs are from Alpha Draconis, a star in the constellation of Draco.

Physical Characteristics: Zeta Dracs are three to five feet tall. They have large round craniums, oversized yellow eyes with pupil slits, ridges for eyebrows, a narrow nose with tiny nostrils, sunken cheeks, and a slit for a mouth. Their light green skin is unblemished and quite smooth.

Belief System: Zeta Dracs believe in service to self. They are taught from the time they are children that humans are the enemy, and that reptilians are superior to all others.

Some Zeta Dracs escaped their reptilian oppressors and became fully conscious. They are now part of the Galacterian Alignment.

Cosmic Agenda: Zeta Dracs are another hybrid race created by the Draconians to expand the tyrannous Orion Empire. They were specifically bred to be controlled and enslaved.

Technology: Zeta Dracs travel with the Draconians in their craft, as their subordinates.

Consciousness Abilities: Zeta Dracs have a special telepathic ability: they are able to psychically enter dark consciousnesses in the lower astral worlds, collect diabolical deeds that have proven successful in the physical realms, and report the evils to the Draconians for further contemplation and execution.

Dimensional Capacity: Zeta Dracs use Draconian travel corridors. Due to the electromagnetic energy enhancements programmed into their auras, they can tune their frequency to become invisible and walk through walls.

Male and Female E'all Reptoids

E'all Reptoids

Universal Origin: E'all reptoids come from Nu Draconis, a double star in the Draco constellation.

Physical Characteristics: Males are six to nine feet and females are five feet ten inches to six feet five inches. Their skin is lizardlike, with colorations from tan to brown, green, and red. They have human-shaped eyes that are brown, gold, or green. Their noses are flat and wide under the bridge, with wide, lipless mouths. This race has varied fingered digits; some have four or five fingers and a thumb, and some are clawed, while others are not. They are muscular and fit.

Belief System: E'all reptoids believe in the Cosmic Law of One. They fled the evil Orion Empire to start their own society, and immediately began obtaining and blending human DNA into their genetics to slowly erase their aggressive behavioral traits, generation by generation. This was done in cooperation with fully conscious humans who understood their plight. The E'all joined the Galacterian Alignment of Space Peoples and Planets about a hundred years ago.

Cosmic Agenda: E'all reptoids are a neutral race and prefer to keep to themselves. They do contribute to the universe by sharing their history with other star nations and younger worlds who are ready to shed their dualistic nature, especially reptilian worlds.

Technology: Although E'all reptoids use Galacterian craft, they prefer their own designs fashioned after birds and eggs.

Consciousness Abilities: E'all reptoids are fully conscious, sovereign beings. They are telepathic and communicate through smell, releasing pheromones that project mental images to the receiver.

Dimensional Capacity: E'all reptoids have multirealm commandership when onboard Galacterian vessels.

Iguanoid Male

Iguanoids

Universal Origin: Iguanoids' original world was in Orion, but they later settled in the Boötes constellation on a water planet.

Physical Characteristics: Iguanoids stand three foot five to five five and use their long tail for balance. Their reptile skin is dark green and tan. They have four extra-long fingers and toes that are clawed. Their orb-shaped eyes are usually black or brown with the common reptile pupil slit.

Another race bred by the Orion Empire, these reptiles are far from the docile iguanas found on Earth. Their DNA blend is half Iguana, a quarter human, and a quarter evil Grey. They were bred with aggressive dominant genes.

Iguanoids wear black hooded cloaks.

Belief System: Iguanoids believe in service to self. They thrive in darkness.

Cosmic Agenda: Iguanoids are troublemakers, conjurors of gloom, and practice the dark arts. They despise the Galacterian Alignment and interfere on worlds that are on the cusp of becoming fully conscious. They primarily target and instill fear in star seeds to prevent them from raising their consciousness. Their goal in life is to see people fail; this feeds their egos and satisfies their need to control others.

Technology: Iguanoids' craft are old saucer technology, decommissioned ships from the Orion Empire.

Consciousness Abilities: Iguanoids have great psychic abilities that infiltrate minds, with the goal being to poison their target, and to lead them astray into dark, fearful places that ultimately imprison them. In short, they steal a person's inner dialogue and make it their own.

Iguanoids feel superior to lesser species and treat them with disdain. But they would never speak out to or against an Alpha Albino Royal Drac or Draconian; they know they would be killed slowly and made example of in a public forum on their world.

Dimensional Capacity: Iguanoids use Draconian travel corridors in the antimatter-universe, black worm-like holes, when trekking between galaxies.

Lizardian People

Universal Origin: Lizardian People are subterranean dwellers from Lacerta, a small northern constellation.

Physical Characteristics: Males and females stand anywhere from five nine to six five. Their torsos have malleable boney plates and their lime-green flesh is patterned after snakeskin. They have the normal slit reptilian eyes, colored yellow, brown, gold, or sometimes green. They have four fingers and a thumb, and five toes. Females run their society, not as Monarchs, but as social stewards towards enlightenment.

Belief System: Lizardian People believe in the Cosmic Law of One. They believe that Source, the heart and soul of Creation, is interconnected to all through the weaving of light, and that creating light in one's life not only brings great joy into your sphere, but to others.

Cosmic Agenda: As members of the Galacterian Alignment, Lizardian People are analytical thinkers and study social behavior in other races. They oversee their negative reptile counterparts, and bring as many of them into the light of consciousness as they can. They want Earth people to know that reptilians and lizard people, or a hybrid of the two, are not all barbaric, and that the negative faction is minimal compared to the whole of their enlightened society. They wish to be recognized as fully conscious beings, awake and aware to the true meaning of life: love.

Technology: Besides Galacterian craft, Lizardian People have tube-shaped motherships and fleet ships.

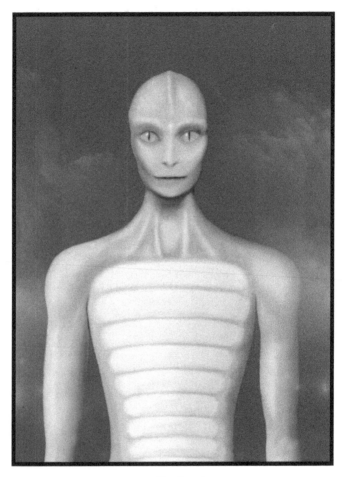

Lizardian Female

Consciousness Abilities: Lizardian People are fully conscious, sovereign, and telepathic beings. They create ethereal music, like whales create songs, and release the melodies on the clairvoyant airwaves to attract a mate, like a telepathic pheromone.

Dimensional Capacity: Lizardian People have multirealm commandership.

Repterrian Male

Repterrians

Universal Origin: Repterrians evolved on Earth and are known as native reptiles. It is said the majority of Repterrians left our world some time ago, joined the Orion Empire, returned, and again thrive in our subterranean worlds.

Physical Characteristics: Repterrian males stand six to eight feet and females are five eight to six feet. Their hides are a blend of dinosaur and the monitor lizard, with colors ranging from tan to brown to green. They have human-shaped eyes with a vertical slit, with vibrant hues of brown, golden yellow, or green. Their noses are flat and wide under the bridge, with large nostrils and wide, lipless mouths. This race has four fat fingers and a thumb with thick human-like fingernails. They are a stalky and muscular group. To heat their cold-blooded bodies, Repterrians have warming rock-beds they lay on underground.

Belief System: Repterrians believe in service to self.

Cosmic Agenda: Repterrians are aligned with the Orion Empire. They manipulate human elites, institutions, and financial systems, influencing religion and militarism—and ages ago they removed human history from historical records and rewrote it to fit into their agenda.

As in all races, not all Repterrians are malevolent. Some have sectioned off to start their own clans with benevolent belief systems on Earth and elsewhere in the universe.

Technology: Repterrians have Draconian disc- or box-shaped craft.

Consciousness Abilities: Repterrians' mental abilities are akin to the Draconians. Through telepathic hypnosis, they can scan the mind of a dualistic being and either try to control their mind or implant subconscious evil deeds they wish for them to carry out, or insert screen memories. They know that if we as a race evolve to the point of becoming fully conscious, they will lose their power over us and they will have to vacate the planet. They cannot live in a thriving positive consciousness based on unconditional love. This is why they do everything in their power to maintain control and keep our world dumbed down, to keep our true nature as sovereign beings out of reach.

Dimensional Capacity: Repterrians use Draconian travel corridors. Like other Draconians, they are also excellent portal skippers. They jump from one location to another, either above or below the planet, or from a planet to one of their vessels in the sky. They built portal technology and tune their energy fields and minds to the tech for instant teleportation.

Reptile Imposter Humans

Universal Origin: Reptile imposter humans are from the Rigel star system in the Orion constellation.

Physical Characteristics: Reptile imposter humans are indistinguishable from Earth humans. They have flawless light to dark brown skin. Women are quite beautiful and men are extremely handsome. They are sexy in a way that presents darkness, mystery, intrigue; masking their true diabolical natures.

Belief System: Reptile imposter humans believe in service to self. They are reptile soul–infused humans of a hybridized caste, bred with dark intention. They are taught how to become more human before infiltrating a human society so they will fit in masterfully.

Cosmic Agenda: Reptile imposter humans are created when Draconian reptiles hybridize with humans over and over in their laboratories until they are human in appearance, and then they are raised to have hateful dispositions, are trained to be highly manipulative, and to worship darkness. These disguised humans are placed on worlds the Draconians wish to infiltrate, like Earth—but empaths and psychically aware humans can sense who they really are. Their subhuman body temperature and blood pressure are lower than normal. And their disguise can easily be revealed by staring into their eyes. It makes them nervous, and their human eyes will shift momentarily to reptile eyes, which can be caught on camera.

Technology: Reptile imposter humans have oval communication devices to talk to their reptile counterparts stationed on or above the Earth. They also visit reptilian motherships periodically to report on their missions.

Reptile Imposter Human Male

Consciousness Abilities: Reptile imposter humans are dark tele-paths that are master magicians at stealing others' thoughts to learn how to manipulate them.

Dimensional Capacity: Reptile imposter humans are third density beings.

Rigelians

Universal Origin: Rigelians are from the Rigel star system in the Orion constellation.

Physical Characteristics: The Rigelian's head is shaped like an upside-down triangle, with large, pointed ears. Its eyes display a full iris with a tiny slit in the middle for a pupil, and clear eyelids similar to a frog. Shoulders are square and taper to a thin waist. They have three extra-long fingers and a thumb with two joints, and three-pronged, pitchfork-shaped feet—all clawed. Their flesh is a milky green.

Belief System: Rigelians are believed to resonate with the Cosmic Law of One.

Cosmic Agenda: Rigelians conduct scientific studies on Earth. They carefully select Earth people to explain their technology to, and share insights about their unique race. Rigelians want their contactees to share this information with the rest of the world.

Technology: Rigelians call their motherships galactic voyagers. They have specialized saucer-shaped craft that are only for inter-planetary travel in certain atmospheres. Their craft have proton ion storage units that are used for rapid accelerative thrust. In the craft engine room, there are two boxes—one is a magnetic fluxerator, and the other is electrostatic. The ship's fuel storage contains free electricity, positrons and positronium, similar to a hydrogen atom. The ships fly through a process of diamagnetism. The metal of the outer craft is made to harmonize with a beam of highly energized photons. The binding energy particle of the atom created is generated outside the nucleus. These particles migrate toward the surface and manifest a magnetic field which repels all matter. The craft can be flown manually or by robots. They have an operating base, a planet in our galaxy.

Rigelian Male

Through a black stone shaped like two triangular bottoms fitted together, worn around their neck, various areas on their craft open by touching the tip of the stone to the wall. The stone is programmed with the key sequence.

Consciousness Abilities: Rigelians are fully conscious and sovereign beings. Their language is a vocalization of chirping, squeaks, and long and short kissing noises. When speaking with other races, both parties wear a headband—the telepathic translator—where wires are inserted into the brain. These wires scan and pick up wave patterns, relaying messages in one language, and then connect to the wave patterns of the other person, passing along the message in their language. Because this is a connected conversation, from brain to brain, the brain stores the entire conversation, and both parties are able to recall the conversation as if being looped on a tape.

Dimensional Capacity: Rigelians are able to travel in the matter and antimatter universe.

Royal Dinoid Female

Royal Dinoids

Universal Origin: Royal Dinoids are from Bellatrix, the third brightest star in the Orion constellation.

Physical Characteristics: Males and female Royal Dinoids stand six to eight feet tall. They are said to be an old universal race, most likely hybridized over the centuries, upgraded to include the best traits and physical attributes of the dinosaur tribes. Their caste is ruled by a monarchy.

Belief System: Royal Dinoids believe in the Cosmic Law of One. They assist those who are ready to become fully conscious, to rise up and out of their former sleeping selves and awaken to the universe and its magnificent kingdom of beings.

Cosmic Agenda: Royal Dinoids are part of the Galacterian Alignment. Their master teachers reawaken their unique culture and past transgressions in their star seeds, who in turn share it with the people of Earth.

However, there are also Dinoid cultures that are negative and aligned with the Orion Empire. These cultures are known as dark dinosauroids, and resemble and carry DNA traits/markers of Tyrannosaurus rex, Utahraptor, Velociraptor, Mapusaurus, Troodon, Spinosaurus, Carcharodontosaurus, and Majungasaurus.

Technology: Royal Dinoids have Galacterian motherships, starships, thought-ships, fleet ships, and scout ships. All their craft and command centers use biological supercomputers. All craft are able to travel through time-space funnels and the innerspace continuum. They do have variations of fleet craft that are shaped like arthropods.

Royal Dinoids have a healing technique unique to star nations: they slip their hands through a strap on gelatinous pads made from liquefied gemstones. Inside the pads' hollow is liquid light, a fluid substance infused with divine consciousness. They pass the pads over the body, and the areas that require attention are healed through the light generated within the pads.

Consciousness Abilities: Royal Dinoids are fully conscious and sovereign beings. Their speech is guttural, but they prefer to use telepathy through pictographs or moving pictures.

Dimensional Capacity: Royal Dinoids live in the fifth dimension vibrating into the sixth.

Saurian Beings

Universal Origin: Saurian beings are from Lacerta, a small northern constellation.

Physical Characteristics: Female Saurian beings are between five and six feet and males are from five nine to six five. While some beaded and scaled lizards are brown, others are a combination of brown, green, and beige, and others are vibrantly colored. Every once in a while an albino lizard is born foretelling of good things to come for the clan. Their reptile eyes are yellow to orange, with one crest running across the middle of their head to join the nose bridge all the way to the nasal openings. Their brows are also ridged. They have semi-small slit mouths and tiny razor-sharp teeth. Their bodies are trim and defined with six clawed fingers and toes.

When a male or female reaches their teenage years, their tail is cut off, and a newer, stronger tail regenerates. When they reach adulthood, it is the individual's choice whether they want to eliminate their tail for good. Those that do cauterize the cut so the tail doesn't grow back. If a tail is kept, it's also used like a steel stand for support, by leaning back on it to take the weight off their feet.

Belief System: Saurian beings believe in the Cosmic Law of One. They live religiously by the All Are One philosophy. Saurian beings are part of the Galacterian Alignment. However, Saurian beings, like all races, have splinter groups in their cultures. Renegade Saurian beings align themselves with the Orion Empire.

Cosmic Agenda: Saurian beings want to educate all star nations about their unique culture. They dabble in genetics, but they are more fascinated by other star nations' cultural similarities and differences.

Saurian Female

Technology: The Saurian beings' basic fleet craft look like crustaceans, and their motherships look like scarab beetles. Glowing rocks on their home world are carved into sacred geometrical shapes and used as healing and warming stones for their bodies.

Consciousness Abilities: Saurian beings are fully conscious and sovereign beings. They are able to transfer smell telepathically when speaking mind to mind—scents that enhance their story.

Dimensional Capacity: Saurian beings have multirealm commandership. Their interdimensional sense of smell locates aromas and pheromones that other Galacterians cannot. When exploring new territories in other dimensions, their heightened sense of smell perceives danger, civilizations, animal and plant life, etc. They say that each realm has its own fragrance.

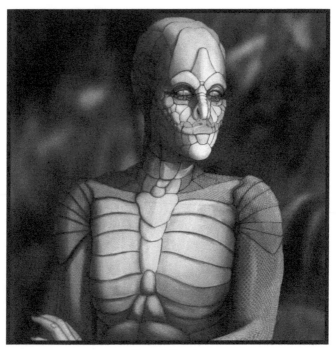

Serpent Female

Serpent Beings

Universal Origin: Serpent beings are from Serpens, a constellation in the northern sky.

Physical Characteristics: Standing about six feet tall, the serpent being's body shape is akin to a human's—but that's where the similarities end. Their vibrant multi-green/brown snakeskin glistens, and they are void of hair. Their eyes are yellow, yellow-orange, or steel silver with slit pupils. They have a nictitating lens, a transparent third eyelid. Their nose is half human and half serpent, their mouth is a slit, and they have a wide serpent tongue.

Belief System: Serpent beings believe in the Cosmic Law of One and are devout practitioners of connecting to Source daily through group meditations.

Cosmic Agenda: Serpent beings may still live underground on Earth. From time to time, they select contactees to share spiritual information with, like the Hopi Indians who named them the Sheti, their snake brothers. As in all races, there are good and bad serpent beings, and renegade offshoots give their caste a bad name.

Technology: Serpent beings enjoy intergalactic travel in elongated fleet and mother craft.

Consciousness Abilities: Serpent beings are fully conscious and sovereign beings. They use telepathic scent pictographs to communicate among each other, an advanced communicative skill derived from using their tongues to collect odors. When communicating with other fully conscious races, they employ telepathic translation devices.

Dimensional Capacity: Serpent beings are interdimensionals. Their tiered consciousness allows them to interact with one another between the layered dimensions.

Small Reptoid Male

Small Reptoids

Universal Origin: Small Reptoids are from Beta Draconis, a binary star and the third brightest star in the northern constellation Draco.

Physical Characteristics: Small Reptoids are about three feet tall and raptor-like—they have a small percentage of snake DNA. Their chests are robust, with muscular arms and springy legs. They have four black clawed fingers and three pronged clawed feet, and are skinned with scutes, scales, and boney plates. Females are brown, beige, and green, while males are brilliantly colored. The females' eye color is brown and sometimes hazel, while the males' are brilliant blue or yellow. Their serpent tongues are longer than most reptiles, and their razor-sharp teeth are lethal. Small Reptoid behavior is aggressive, relentless, and annoying. The Alpha Albino Royal Dracs and Draconians have been known to kill them when they get on their nerves.

Belief System: Small Reptoids believe in service to self and are aligned to the Orion Empire.

Cosmic Agenda: Small Reptoids were genetically created by the Draconians. They're Orion's military scouts, designed for their swiftness, agility, and keen eyesight. They act on instinct, not on morals. During battles, it is their speed that produces chaos, ricocheting off the enemy like pinballs, dropping them, digging their poisonous claws and teeth into flesh until the target ultimately succumbs.

Technology: Small Reptoids only travel in Draconian vessels, ships that are disc- or box-shaped and echo with a dark sonance. They are allowed to use sonic blasters, a Draconian invention, during battle. It destroys a person's aura, their life force, making them weak and defenseless.

Consciousness Abilities: The Draconians have the Small Reptoids' brains linked to a device that controls them. The Small Reptoids are telepathic in the sense that they send pictures to others in the tribe when danger might present itself. Otherwise, they have a hissing style of verbal speech with short chirps.

Dimensional Capacity: Small Reptoids use Draconian travel corridors. Because Small Reptoids only travel in Draconian craft, they become very excitable and unmanageable when sliding between dimensions, so they are sequestered to their own compartment.

Part Six

OTHER

EXTRATERRESTRIALS

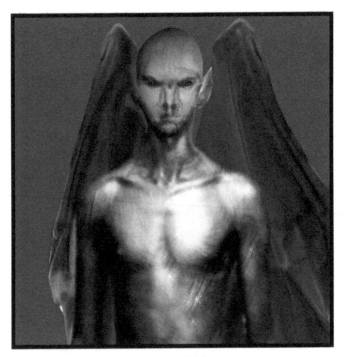

Bat Bowoul Male

Bat Bowouls

Universal Origin: We don't know where Bat Bowouls come from.

Physical Characteristics: Bat Bowouls are winged, short, stubby, and stand anywhere from two to four feet. Their bat faces are hideous: they have razor-sharp teeth; slimy, wide noses; and mesmerizing black eyes. Their black-and-white flesh is cold and clammy and similar to frog skin.

Belief System: Their religion is dark, sadistic, and demonic. They only serve themselves.

Cosmic Agenda: Bat Bowouls are enslavers of anger, contempt, distraction, depression, despair, envy, grief, jealousy, pain, pensiveness, suicidal urges, and suffering. They want to destroy as many star seed souls as they can. They abhor the Galacterian Alignment and their consciousness-raising program.

Technology: Bat Bowouls' technology capabilities are unknown.

Consciousness Abilities: Bat Bowouls employ telepathic hypnosis that manipulates the lower part of the brain and forces the subject to be compliant. They feel their victims' fear, gain control over them, and destroy their physical bodies and souls through substance abuse.

You can thwart Bat Bowouls from mind control by raising your consciousness through spiritual practice, meditation, connection to Source, and shielding your mind and multidimensional bodies with a self-imagined shield of Archangel Michael.

Dimensional Capacity: Bat Bowouls live in the dark astral realm. They slither through this world and choose their victims to destroy, and enter their minds from behind the veil. Only a strong psychic or empath is able to detect them.

Jowlen Male

The Jowlen

Universal Origin: The Jowlen come from the Eridanus constellation, the sixth-longest constellation in the sky.

Physical Characteristics: The Jowlen's short, stubby body is anywhere from two to four feet. They waddle when they walk, yet are nimble. With wrinkled flesh and overlapping folds of skin, their pigmentation ranges from dark blue to cobalt blue, white or tan. They have four stubby fingers and a thumb, with matching numbered toes. Hooded cloaks are their primary attire. Their human eyes are pale blue, brown, or black.

Belief System: The Jowlen believe in the Cosmic Law of One. They are wise sages.

Cosmic Agenda: The Jowlen are members of the Galacterian Alignment. Observed in abduction scenarios and in benevolent visitations, these beings are seen as medical and/or spiritual doctors. They assist in genetic missions of other extraterrestrials. When star seeds' mental faculties and belief systems elevate, their frame of mind fights these new spiritual universal ideals; the Jowlen have been known to assist in bringing them back to a healthy equilibrium.

Technology: The Jowlen are able to move around easily by wearing a levitation device. They also employ various handheld healing tools.

Consciousness Abilities: The Jowlen are fully conscious, sovereign, and telepathic beings.

Dimensional Capacity: Jowlen are multidimensional. Due to the electromagnetic energy enhancements programmed into their auras, they can tune their frequency to become invisible and walk through walls.

Unknown interdimensional alien being (close-up detail at left)

Unknown Alien Caught on Camera

Universal Origin: Because I am the photographer of this alien being, my first thought is that it's from the Serpens constellation, possibly of serpent, reptile, and/or Saurian descent. On June 2, 2017, I was researching an abduction case involving alien visitors and snapped a succession of shots. The first shot was of a being in shadow that was not present at the time the photo was taken. The woman whose back is to us in the forefront has at her waist a light source with human fingers coming out of the light. The strip of light below her jacket is her white shirt illuminated from my camera flash. I also rendered a lightened version of the alien's face in close-up to show more detail. The shot I took right after the alien shows streams of white plasma falling from the sky.

Physical Characteristics: This alien was about six three. It had an extra-long, stalky torso, an elongated neck, and short legs. When the photo is lightened, the alien is wearing a one-piece spacesuit and eye goggles perhaps to see interdimensionally like we might use night vision binoculars. It's holding a round light source, which I believe to be an interdimensional device.

Belief System: The alien's belief system is not known.

Cosmic Agenda: The feeling I get when looking at this being is that it's a neutral race, even though its appearance might frighten most.

Technology: The alien's technology was interdimensional.

Consciousness Abilities: If this alien is able to launch himself from one dimension to another undetected, his race are most likely strong telepaths, and possess advanced mental faculties.

Dimensional Capacity: The alien was multidimensional.

Top: Lightened version of alien's face
Bottom: Wisps of plasma falling from sky

The Galacterian Alignment of Space Peoples and Planets

The **Galacterian Alignment of Space Peoples and Planets** is a universal alliance of fully conscious beings that have united their planets to benefit all universal-kind. They work with the University of Melchizedek, an outer space city that has 490 learning spheres in the Mira System. Commanders and sub-commanders are trained at Melchizedek. The highest ranking commanders are Solar 1, 2, and 3 commanders. Secondary are the sub-commanders 1 through 8. A sub-commander of the 1st rank is one step away from becoming a Solar 3 commander. Solar 1s are the highest ranking.

The **Star Seed Alignment** is a subdivision of the Galacterian Alignment. It's one of the many established consciousness-raising programs. Universal citizens temporarily leave their lives in the stars to reincarnate on worlds in the pangs of duality, such as Earth. A portion of the fully conscious being's soul will incarnate while their body and remaining soul hibernates. Alignment protocol facilitates the process: as the soul evolves, raising its consciousness, a little more of the original soul joins itself on Earth, because the body and mind calibrate to its raising frequencies. This helps elevate spiritual consciousness not only for themselves but for others around them. Generations down the line, star seeds begin raising star seed children. Because of the generational genetics, more of the original soul is able to enter an Earth form and advance at an accelerated rate. Some kid and teen star seeds today are already drawing star maps and speaking five or more star languages. The ultimate goal for groups of star seeds is to merge their duality and become fully conscious, and then the rest of the planet follows.

Star nations of the Galacterian Alignment share a common set of technology and healing tools, they have a code of ethics, and they refrain from using any kind of monetary system. They adhere to the Cosmic Law of One, live in a unified field of trust, and love everyone equally. Although their craft are similar, some races have their own tech married to the main tech. Ships may work on different principles or have a unique design based on individual preference. They have a saying, "Everything was created out of love and respect and therefore is performed in original intent."

Fleet Ships

Fleet ships are designed to perform many functions, including planetary and interplanetary travel, time travel, and thought-jumps. Fleet ships plot a course from their point of origin and arrive at their destination by traveling on a charted navigational beam. Think of it as a monorail in outer space, with the beam being the track and the ship being the monorail.

Fleet ships have ultra-advanced force fields. They are equipped with deflection weaponry and all of the latest spiritual technologies. Some are defense craft.

The crew capacity is one hundred, two hundred, or three hundred, depending on the mission of the ship. Star nations have many ship models, but most are saucer-shaped. Others shapes include boomerang, egg, domed, flattened with bottom-and-top-dome protrusions, and triangular.

Space is not an issue onboard any vessel; the Galacterian worlds use contraction and expansion tech, based on the same contraction and expansion principles of the universe. For instance, a small egg-shaped craft from the outside looks like it might fit two people, but once inside, there is an appearance of infinity; the laws of physics are obliterated.

All craft are able to hide in a pocket dimension they create to cloak, or in another dimension to monitor this dimension from real-time screens.

The fleet is linked to the data banks of the Living Library of Akasha, where all life is recorded (the soul being the recorder) and stored, and can be accessed at any given moment.

Fleet ships are just as agile in the water as in the air, in complete control of the elements. Upon entry into any body of water, a slippery cocoon of light surrounds the craft, its protective barrier allowing the ship to slide through liquid with ease. It's also designed not to disturb ocean life.

The fleets are equipped for scientific exploration of our multidimensional universe and other universes.

Merkabah Vehicles

The elders, created beings, and Angelic Corps travel in Merkabah vehicles, a transport of Living Light (luminous briolette star vehicles) they create around themselves to either go short distances, or enter space portals to other sectors of the universe.

Motherships

The most common shape for motherships is spherical. Other designs resemble a giant oval or a traveling country. Colossal mother craft can house seven million or more. There are science labs, organic consciousness engineering labs, command centers, science centers, spiritual centers, star centers, spiritual technology stations, farms, outdoor environments, and creative arts environments.

Motherships have ultra-advanced force fields. They are equipped with every advantage in space exploration, defense and deflection weaponry, and all of the latest spiritual technologies.

From their point of origin, motherships plot a course to their destination, and then ride a navigational beam upon entering the innerspace continuum. Think of it as a monorail in outer space, with the beam being the track and the mothership being the monorail.

Motherships are able to hide in a pocket dimension they create to cloak, or in another dimension to monitor this dimension from real-time screens.

Motherships are linked to the data banks of the Living Library of Akasha, where all life is recorded (the soul being the recorder) and stored, and can be accessed at any given moment.

If you're a visitor from another world in the multiverse, the mothership automatically scans your biological and spiritual bodies to tend to your every need. For instance, if you require a special type of oxygen, it will create a life support system around you in the form of an invisible cocoon. From food, to shifting densities, to anything that might be required, all is taken care of.

Motherships are multidimensional realm vessels.

Plasma Craft

Plasma craft, a very advanced technology, are a marriage of organized and rarefied plasma compared to intracluster mediums, high vibratory metal, and organic materials. The craft are living divine beings at the avatar level. When in flight, they transform all dense matter into its natural spiritual state to jump across the universe with a mere thought, navigate the impossible, split its structure into separate parts and reunify, hide in other or temporarily created dimensions, all while using an impenetrable ultra-advanced force field. These craft are equipped with every advantage in space exploration, deflection weaponry, and all of the latest spiritual technologies.

The craft are linked to the data banks of the Living Library of Akasha, where all life is recorded (the soul being the recorder) and stored, and can be accessed at any given moment.

Plasma craft are similar to thought-ships, but unique in the sense they are powered by infusion of the creational light force.

Plasma craft are just as agile in water as they are in the air.

All plasma craft are multidimensional realm vessels.

Scout Ships

Scout ships are small craft that gather environmental information on explorative worlds and inside cosmic bodies. They have the capabilities to map, measure, and take samples. Other scout ships are larger and usually hold up to three to eight people for scientific excursions. Orb- or pod-shaped are the most common designs.

Aquatic scout ships are just as agile in water as they are in the air.

Specialty scout ships can travel into unknown multiverses and gather information.

Starships

Starships are smaller than motherships but have the same amenities. These vessels are designed like rods (smaller rod craft are supply ships), revered birds, or discuses, or are modeled after geometrical patterns.

Starships have ultra-advanced force fields. They are equipped with every advantage in space exploration, defense and deflection weaponry, and all of the latest spiritual technologies.

All craft are able to hide in a pocket dimension they create to cloak, or in another dimension to monitor this dimension from real-time screens.

Starships are linked to the data banks of the Living Library of Akasha, where all life is recorded and stored, and can be accessed at any given moment. Starship vessels are heavily equipped for scientific exploration.

The majority of starships are multidimensional realm vessels.

Thought-Ships

Specialized motherships and thought-ships can jump across the seven superuniverses with a mere thought with hierarchal permission. A commander's DNA is infused into the biological metal, organic brain, and organs of their personal organic craft—their thought-ship. This enables them to travel across universes and superuniverses. The ship becomes a clone of the commander, an extension of their body, and they are the only one who can operate it—a Galacterian Alignment precautionary measure. Everything created has a resonance signature. The commander of the vessel tunes the resonation of the craft to the destination's resonance signature, and the ship and destination become symbiotic—thus, instant teleportation. Here one instant, there the next. Each ship's personality is tended to on a physical, mental, and spiritual level. Although the intelligence of the craft is part of the commander, the organic AI has its own personality and is capable of reasoning, learning, and self-improvement. Navigational commanders' DNA is also infused into motherships' biological metal, organic brain, and organs.

If a thought-ship is captured or destroyed, it rapidly disintegrates, because of the technology written in the metal. This guarantees that negative alien races or civilizations not ready for advanced knowledge will never gain the spiritually advanced technology. Each craft has scout ships that can be unmanned or house up to three persons.

Thought-ships are able to hide in a pocket dimension they create to cloak, or in another dimension to monitor this dimension from real-time screens.

Thought-ships are linked to the data banks of the Living Library of Akasha, where all life is recorded (the soul being the recorder) and stored, and all information can be accessed at any given moment.

Thought-ships are just as agile in the water as in the air, in complete control of the elements. They are equipped for basic scientific exploration.

Thought-ships are multidimensional realm vessels.

Cosmic Law

When God spoke the covenant to Moses, delivering the Ten Commandments, those moral laws were intended for mortals. The Comic Law of One is written for fully conscious beings, created beings, angels of all classes, etc. They are the laws of harmony; the inherent laws of nature; the inclusion law of all, of creational imaging, of equality, of intelligence throughout the body of creation.

Cosmic Law of One: These laws are designed to align lives to the universal frequency of harmony.

1. Law of Unity: Everything is interconnected. When you are at one with yourself, you are at one with everyone. There is no separation.

2. Law of Vibration: Tuning the spiritual body to vibrate at a high frequency enriches life.

3. Law of Manifestation: Using your frequency to manifest your hopes, dreams, and desires.

4. Law of Cause and Effect: The energy/frequency you put out is the energy/frequency you get back.

5. Law of Mental Clarity: Connecting to Source daily for mental clarity.

6. Law of Spiritual Development: Connecting to Source daily to design your spiritual path towards Divinity.

7. Law of Physical Health: Eat right, exercise, and laugh.

8. Law of Social Interaction: Maintaining close relationships with family, friends and making new friends does the heart good.

9. Law of Compensation: Working towards personal goals and achieving the end result no matter how challenging the path might be.

The Contrary Cosmic Law of One: This was written for the fallen angels and for races and individuals who disobey the natural laws, and have branched away from their divinity, and now covet service to self. When these laws are broken, there is a trial, and those sentenced are incarcerated on one of the seven prison planets (dependent of severity of crime) in the Ghost Galaxy located in the Satania System. For heinous crimes without remorse, a lengthy period of time is designated for redemption discovery. If redemption is sought, rehabilitation follows. If remorse is never found in their hearts, a new trial will determine their soul's fate; either they will live out their days on a prison world, or be sentenced to soul-death. Some of the laws are as follows:

1. Thou shall not kill; for you kill a part of yourself, for every being carries the blueprint of the primordial atom.

2. Thou shall not enslave, imprison, kidnap, or hold anyone hostage.

3. Thou shall not mentally or physically harm or torture another.

4. Thou shall not sexually violate another.

5. Thou shall not psychically attack another.

6. Thou shall not invade the private thoughts of another.

7. Thou shall not hinder the free will of another and their right to choose.

8. Thou shall not create life forms from genetic materials of your own or another's race, unless sanctioned by the Angelic Corp.

9. Thou shall not abduct cosmic citizens against their will to perform medical tests or to extract genetic materials.

10. Thou shall not alter the biological chemistry of another without their consent.

11. Thou shall not remove or lock the ascension chakras of another.

12. Thou shall not interfere in an individual's or races' evolution and impede their right to ascend.

Dark Undermonic Laws: The laws of service to self; to gain power, technology, and material wealth. They defile all that is good and righteous, and mock the Cosmic Law of One. These undermonic laws are the opposite of the harmonic laws. They are the laws of defilement of God, deceit, divide and conquer, robbing younger civilizations of their genetic materials, creating aggressive hybrid races for nefarious purposes, energy siphoning, hate, indulgence, perpetual sin, self-gratification at others' expense, and vengeance.

APPENDIX C

Otherwordly Terminology

Ancients of Days: Paradise Sons of previous universal programs; the Triune rulers of the seven superuniverses. (See also paradise sons.)

Constellation Fathers: The Creator Gods who oversee their constellation jurisdiction.

Dualistic Being: Two opposing concepts: right vs. wrong; good vs. evil; service to self vs. service to humanity. When a person merges their dualistic nature, they understand why the pulses of both concepts were created: to evolve the soul. Once becoming fully conscious, utilizing 100 percent of the brain in union with their recent conscious linking to Source, life is then lived in harmony, using the heart to rule the mind because emotions have been mastered. Diversity is a spiritual tool designed to perfect humans.

Eternals of Days: These Paradise Sons are equal to the Ancients of Days; only the Eternals oversee one billion worlds of Havona, the center of all Creation, the sublime perfection of Paradise. The Godhead created exactly one billion Eternals, one to oversee each Havona world. (See also ancients of days, Havona and paradise sons.)

Fully Conscious Beings: A term used when a soul has merged its dualistic nature (light and dark) to work in harmony for the whole of universal-kind instead of the self. It is the unification of the heart and mind, mastery over emotions, moving from a seven base chakra system to twelve, attaining full divine power.

Galactic Command Centers: These underground facilities on Earth are constructed and utilized by the Galacterian Alignment. From these outposts they monitor the world, including its leaders and aggressive citizens (this and otherworldly) of interest. Other command centers are onboard motherships, starships, and fleet craft.

Havona: The Eternal Paradise that consists of one billion worlds. The visual intake is beyond anything seen on a world such as Earth. (See also eternals of days.)

Light Essence Language/Language of Light: An advanced methodology of communication in Oneness. This frequency is a sacred communion between enlightened beings. It may also include geometrical designs and sacred codes. It's an energetic language of the heart, not the mind.

Living Light: Light infused with creational consciousness and life force.

Merkabah Star Vehicle: A three-dimensional tetrahedron; a light vehicle that enshrines the body by the will of the programmer and is powered by their thoughts.

Multirealm Commandership: Specialized Galacterian craft are able to physically exist in every dimension at the same time. The crew in each dimension belongs to that realm; although the members are able to disassemble and reassemble their bodies in any dimension on the ship and interact with those crew members and craft. The multirealm commanders' daily exchange of universal information is vital to maintain proper equilibrium between dimensions. What happens in one dimension affects the other layers subtly.

Orion Empire: Owned by the malevolent Draconian reptiles. The Draconians are quite intelligent and master manipulators of cosmic law. Their plan is to own every planet in the universe, not in the usual sense of seize and control, but to take each from within and rule it in secret for all its resources.

Oversoul: The soul in its perfection form that is unified with creation. Once an oversoul has gleaned much incarnation experience and is at an advanced state, it can split into seven embodiments simultaneously. These lifetimes can be spread throughout the universe on various planets. They are deemed Mortal Finalitiers. Some extremely advanced souls have been granted the abilitiy to experience a lifetime in each of the seven superuniverses. They are deemed Eternal Finalitiers.

Paradise Sons: God the Father, or our Paradise Father, creates Himself in many forms of Himself. These sons run His spiritual network across the seven superuniverses. The Father is in the Son and the Son is in the Father. Each Son is unique and has a separate personality. Paradise Sons of previous universal programs are known as the Ancients of Days.

Seven Superuniverses: Within each superuniverse there are 700,000 local universes. Superuniverse one: Flowninington; superuniverse two: Nottington; superuniverse three: Blennington; superuniverse four: Sonhueton; superuniverse five: Shreeton; superuniverse six: Evantanton; superuniverse seven: Orvonton. Earth's registry number in this superuniverse of inhabited planets is 5,342,482,337,666.

Solar Consciousness: Higher dimensional consciousness. The wisdom, knowledge and understanding of the Living Light and its enlightenment codes that create the multiverses. The keys to life and existence.

Soul-Death: Life-sentenced universal prisoners with no remorse are given a final judgment at a trial overseen by the Ancients of Days. If there is no hope that the soul will ever change, they are sentenced to soul-death, or absorption back into the heart and soul of creation. This ends the lineage of that soul forever.

Soul Vector Suit: A genetically engineered body married to the host's DNA and timestamped consciousness. This temporary body is utilized when visiting other worlds, either because their current body cannot endure traveling a lengthy universal distance or because of an environmental sustainability issue. The body resembles the actual person; only the skin is a fluorescent milky-white. Nonhuman terrestrials when visiting a human world on a cultural exchange program for instance, may inhabit a human vector soul suit and vice versa. Or in the case of the Digital Immortals, their soul vector bodysuit is engineered for permanence.

Spiritual Technology: Biological technology created to work in harmony with cosmic law.

Star Seeds: (See also star seed alignment in appendix a.)

Synchrotron Sky: An infinite-appearing sky that synchronizes day into night and regulates temperature. This technology is used in outdoor settings on various floors onboard craft and in civilizations that live on the interior of their worlds.

Time-Space Funnels: A temporary funnel comprised of a liquid and light substance created with Alignment technology. It's somewhat similar to a wormhole, except it's much shorter in length; and, it's used to launch vessels into and out of the innerspace continuum.

Union of Days: The Union of Days are Paradise Counselors and advise evolving local universes. Their duties are to monitor the reflecting thoughts of the hierarchy and other races in question. This cumulative monitoring is known as reflectivating. This monitoring is not an invasion of private thoughts; it is more like an intuition that the universal official or race in question is in need of spiritual guidance away from degenerative thoughts that might lead to the birth of the ego.

Yod Spectrum: The newly fully conscious being's twelve DNA strands or chakra system is activated and now connects the individual to advanced dimensions and to their universal brethren for further edification on the cosmic courses of life. The physical body is now able to reconstruct itself cell by cell with Living Light to transform itself into a Divine Body. Divine purpose is also activated.

FURTHER READING

To learn more about the Agarthans go to the Facebook page of Commander Lady Athena. Also read *The Smoky God* by Willis George Emerson.

To learn more about the Angelic Corps, the I AM THAT I AM teachings, the Melchizedeks, the Moramiams, the Pleiadians, and the Seventh Ray Race, read *The Autobiography of an ExtraTerrestrial Saga* book series. *www.AutobiographyOfAnET.com*

To learn more about the Apunians, read *Contact from Planet Apu: Beings from the Future Among Us* and *The Ark: An Extraterrestrial Warning from Alpha Centauri*, by Ricardo Gonzalez. Legado Cosmico; *www.legadocosmico.com*

To learn more about the Arcturians, go to Viviane Chauvet's website. *www.InfiniteHealingFromTheStars.com*

To learn more about the Arians and the Gne, read *UFO Contact from Arian of Aldebaran* by Wendelle C. Stevens and Martin Wiesengruen. *www.UFOPhotoArchives.com*

To learn more about the Clarions, read *Beyond the Heavens: A Story of Contact* by Maurizio Cavallo. *www.centroclarion.it; www.facebook.com/mauriziocavallo.jhlos; www.facebook.com/Jhlos*

To learn more about the Iargas, read *UFO Contact from Planet Iarga* by Stefan Denaerde, contactee, and Lt. Col. Wendelle C. Stevens (Ret.). *www.UFOPhotoArchives.com*

To learn more about the Itipurians, read *UFO Contact from Planet Itibi-Ra: Cancer Planet Mission* by Ludwig F. Pallman and Wendelle C. Stevens. *www.UFOPhotoArchives.com*

To learn more about the Klermers, read *UFO Contact from Planet Klermer: Possibilities of the Infinite* by Rodolfo R. Casellato, M. A. O. Bianca, and Wendelle C. Stevens. (*www.UFOPhotoArchives.com*).

To learn more about the Koldashans, read *UFO Contact from Koldas: A Cosmic Dialogue* by J. Carl van Vlierden and Wendelle C. Stevens. *www.UFOPhotoArchives.com*

To learn more about the Lady of Light, explore interviews with Christopher Bledsoe Sr. and the Fayetteville Encounter.

To learn more about Mantis Beings and the light language, go to *www.JacquelinSmith.com.*

To learn more about the Mothman, read *The Mothman Prophecies*—or see the movie starring Richard Gere and Laura Linney.

To learn more about the Original Human Orions and Sirius-Solar-Empowerment (Blue Sirians) and the Guardians, go to Lightstar's YouTube page. *www.LightStarCreations.com.*

To learn more about the Rigelians and the Guardians, read *UFO Contact from Beyond Rigel: A Cherokee Girl's Odyssey* by Lucias Farish, Wendelle Stevens, and Phyl Pierceall. *www.UFOPhotoArchives. com*

To learn more about Saint Germaine, Guy W. Ballard, and the "I AM" Teachings, go to *www.saintgermainfoundation.org.*

To learn more about the Sassani race, go to *www.Bashar.org.*

To learn more about *Stranger at the Pentagon* and Valiant Thor, go to *www.StrangerAtThePentagon.com.*

To learn more about the Ummites, read *UFO Contact from Planet Ummo: The Mystery of Ummo* (Volume 1) by Antonio Ribera, with contributions by Ignacio Darnaude Rojas-Marcos and Wendelle C. Stevens and edited by Cece Stevens. *www.UFOPhotoArchives.com*

To learn more about the beings from Zeta Reticuli, read *Captured! The Betty and Barney Hill UFO Experience* by Stanton T. Friedman, MSc., and Kathleen Marden. *www.Kathleen-Marden.com*

IMAGE CREDITS

Agarthan (Saint Germaine), Antarians, Twenty-Four Elders of Andromeda, The Angelic Corps (Archangel Michael), Altarian, Arians, Cassiopeians, Ceitans, Celestials, Proxima Centaurians, Cyclops, Cygnus Alphans, Eridaneans, Itipurians, Klermers, Lyrans, the Melchizedeks (Father Melchizedek and Machiventa Melchizedek), Alcyone Pleiadians, Pleiadians, Renegade Pleiadians, Sagittarians, Human Sirians, Solar Light Beings, Superangels (created superangel Gavalia and ascendant superangel Galantia), Vegans, Venusians, Watchers, Moramiams, Zeta Reticulans, Beesonites, Seventh Ray Race, Draconians (Dracs), Dragon Dracs, Reptile Imposter Humans, The Jowlen, and Bat Bowouls artist Christine Kesara Dennett © Craig Campobasso. Craig Campobasso and Christopher Bledsoe Sr. photograph and Unknown Alien Photographs © Craig Campobasso. *www.AutobiographyOfAnET.com*

Albert K. Bender taken by retired AF photographer August C. Roberts. Alpha Albino Royal Dracs and Dinosaurian artwork © Eric Lutes. *www.EricLutesArt.com.* Apunian artwork © Ramiro Rossi. Blog: *ramirorossioficial.blogspot.com.* Facebook: Ramiro Rossi Artista Visual. Instagram: @ramirorossioficial. Andromeda-Healing-Matrix, Arcturus InfinityMind, Orion Magical Creation, Sirius-Solar-Empowerment, Guardian Sacred Builders (created for Ryk Hall) artwork © Lightstar, Lightstar Creations. *www.LightStarCreations.com.* Clarion Photographs © Maurizio Cavallo Jhlos. *www.centroclarion.it.* Craig Campobasso photograph by David LaPorte. Frederick Warner Vierow photograph © John Palean. Iargan man, and twelve dimensional synchronization machine artwork by Rudolph Dass. To see the full rendition of "The Lady of Light" in color (oil on canvas 66 X 35) and more of Doug Auld's art, go to *www.DougAuld.com.* Lion-Feline, Royal Dinoids, and Iguanoids artwork © Paige Benson. Find more of Paige's art at *www.wattpad.com* @ShadowMelody284. Lizardian People, Saurian Beings, and Zeta Dracs artwork © David Chace. *www.facebook.com/davidwchace.* Procyon, Titan Sirian, Sirian Insectoid, Soulzars, and Ummites artwork © Kim Edward Black. *www.KimEdwardBlack.com.* Zeta Reticuli bust photograph of Junior from the Betty and Barney Hill case courtesy © Kathleen Marden. Used with permission. *www.Kathleen-Marden.com.* Rigelians and Serpent Being artwork: Valoia Laolagi. *www.Artstation.com/valoia.* Sassani images from the documentary film *First Contact* Copyright © 2016–2020 by Zia Films by LLC. *www.ziafilms.com.* Used with permission. Valiant Thor (Created Being) taken by retired AF photographer August C. Roberts. For more info on Valiant Thor, go to *www.StrangerAtThePentagon.com.* Zeta Human by Leo Blanchette. *clipartillustration.com*

Images by Shutterstock: Anunnaki by Kamira Stone; Clones by Ana Aguirre; Digital Immortals by Bruce Rolff; Men in Black by delcarmat; Synthetics by GarryKillian; EBEs by Pavel Chagochkin; Short Greys and Orons by Adike; Tall Greys by Sciepro; Taller Greys by Sarah Holmlund; Zeta Humans—Blue and Golden Caste by First Step Studio; Mothman by Esteban De Armas; Mantis Beings by Ralwell; Grey Insectoids by Design Projects; Zeta Lizard Humans by Joe Prachatree and Zeta Lizard Human skin by Semmick Photo; E'All Reptoids and Repterrians by Patrick Ruzic; Small Reptoids by Katja Gerasimova.

Also by Craig Campobasso

The Autobiography of an ExtraTerrestrial Saga
BOOK SERIES

Book One: I AM Thyron

Book Two: Waking Thyron

Book Three: Thyron's Dossier

Book Four: The Huroid Revolution: And Other Warring Creatures

The Silence of the Hams: A Pictorial Memoir of the Making of a Cult Classic

ABOUT THE AUTHOR

Award-winning filmmaker and Emmy-nominated casting director Craig Campobasso was fifteen when he started in the entertainment business. After graduating high school at age seventeen, Craig went to work behind the scenes on Frank Herbert's *Dune*, directed by David Lynch, and two Arnold Schwarzenegger movies: *Conan the Destroyer* and *Total Recall*. He began his casting career on Steven Spielberg's *Amazing Stories*, and later received an Emmy nomination for Outstanding Casting for a Series on David E. Kelley's *Picket Fences*. Craig's casting career spans three decades.

Craig's mother, Marie Donna King Campobasso, told him that she knew when he was still in the womb that he would become a writer. He fulfilled that prophecy at age twenty-six, after experiencing a life-changing spiritual awakening, and *The Autobiography of an ExtraTerrrestrial Saga* series was born. His passion is to write stories that provoke readers to think, raise their consciousness, and expand their minds about Creation, while entertaining in the Hollywood tradition.

Craig wrote, produced, and directed the short film *Stranger at the Pentagon*, which was adapted from the popular UFO book authored by the late Dr. Frank E. Stranges. The film won Best Sci-Fi Film at the Burbank International Film Festival and a Remi Award at the Worldfest Houston International Film Festival.

Craig has appeared on *Coast to Coast AM* and *Beyond Belief,* hosted by George Noory. He has also appeared on *Ancient Aliens*, where Giorgio A. Tsoukalos is the main ancient astronaut theorist.

Craig's series, *The Autobiography of an ExtraTerrestrial Saga*, will expand to seven volumes. He is currently working on book five and is preparing the feature-length film *Stranger at the Pentagon*.

To learn more, visit:
www.AutobiographyOfAnET.com; www.StrangerAtThePentagon.com;
and *www.CraigCampobasso.com*